THE STUDY OF SOIL
IN THE FIELD

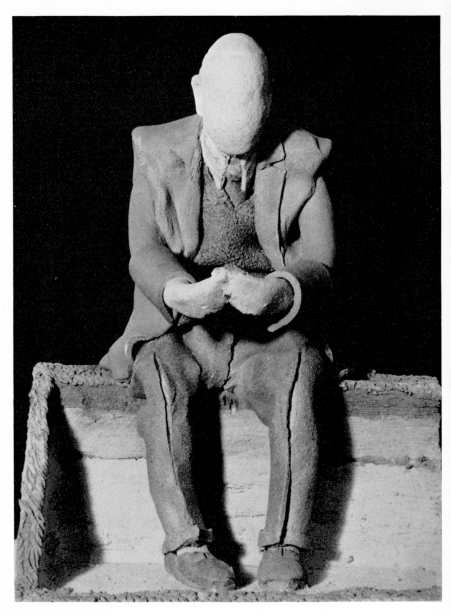

CLAY STATUETTE OF THE AUTHOR BY
AUDREY BLACKMAN

THE STUDY OF SOIL
IN THE FIELD

G. R. CLARKE, B.Sc., M.A.

SOMETIME DEMONSTRATOR IN RURAL ECONOMY
AND LECTURER IN SOIL SCIENCE IN THE UNIVERSITY OF OXFORD

ASSISTED BY

PHILIP BECKETT

LECTURER IN SOIL SCIENCE IN THE UNIVERSITY OF OXFORD

FIFTH EDITION

CLARENDON PRESS · OXFORD

1971

Oxford University Press, Ely House, London W.1

GLASGOW NEW YORK TORONTO MELBOURNE WELLINGTON
CAPE TOWN SALISBURY IBADAN NAIROBI DAR ES SALAAM LUSAKA ADDIS ABABA
BOMBAY CALCUTTA MADRAS KARACHI LAHORE DACCA
KUALA LUMPUR SINGAPORE HONG KONG TOKYO

FIRST EDITION 1936
SECOND EDITION 1938
THIRD EDITION 1941
FOURTH EDITION 1957
FIFTH EDITION 1971

PRINTED IN GREAT BRITAIN
AT THE PITMAN PRESS, BATH, SOMERSET

Preface
to the Fifth Edition

ROBIN CLARKE retired in 1962, to live at Adderbury on the Cotswolds. There he has written the new sections for this edition, the mellow fruit of sixty years' observation of the soil, developed by accretion on the little handbook which was the first edition in 1936. This is still his book and a very proper crop from the Banbury silt loam, deep phase, which (Robin taught us) is the ideal to which all other soils aspire, and on which inevitably he now lives.

Born in 1894, G. R. Clarke was brought up in Oxford. From 1909, shortly after the foundation of the Department of Rural Economy, until he joined the Army in 1914, he was laboratory assistant to C. G. T. Morison, later Reader in Soil Science. He returned to Oriel College in 1919 to read chemistry under Tizard and Hammick, taking his B.A. in 1921 and a B.Sc. on soil nitrogen under Morison in 1923. He was appointed advisory soil chemist for the counties of Oxfordshire and Northants in 1924, and returned to Oxford as Demonstrator in Soil Science in 1930.

Meanwhile he had attended the Third International Congress of Soil Science in Leningrad. He returned a convinced pedologist, and his impressions from that visit coloured all his subsequent work: the Russian influence is very apparent in this book.

Robin Clarke brought back some monoliths from that excursion and soon set about persuading delegates to the next Congress (Oxford 1935) to bring or send monoliths of their favourite soils. Many of these are still in Oxford, and the collection grew through the years, by gifts from visitors and former pupils, until it contained over 150 monoliths from all corners of the world on Robin's retirement.

The study of the soil in the field was written shortly after the 1935 congress, with overseas students particularly in mind, at a time when pedologists were still groping for a uniform terminology. As C. G. T. Morison wrote in his preface to the first edition:

The study of the soil in the field makes demands upon the worker of a somewhat peculiar nature: it requires not merely a knowledge of natural science of no mean order, but it demands a faculty for keen and accurate observation of details which are by no means easy of observation, and which may easily escape the notice of any but the most practised eye.

Only if observations of this kind are to some extent codified do they become exchangeable between one person and another.

Mr. Clarke helped much by his sympathetic feeling for this aspect of soil work, has made it his special study, and has endeavoured in this book to put together his own experience and that of other workers in the same field in such a way as to be of assistance to students and others.

This has remained the aim of all subsequent editions, including this one. Generations of students have sat round soil pits with Robin, feet dangling, absorbing soil lore after beer and sandwiches in a well-chosen pub: they will recognize the familiar flavour. I have done no more than prune some redundancies and see the work through the press. This I have been glad to do for old friendship, and because it now gives me the opportunity to thank Robin for much forbearance and counsel from the time when I first joined him in soil science in 1950.

PHILIP BECKETT

Oxford, 1970

Acknowledgements

I WISH first to thank all those friends whose help and advice have always been so kindly given to me during the past sixty years. Many of them are now dead and among these I must particularly mention C. G. T. Morison under whose instruction I began to study soil in 1909. More recently I have been aided by B. W. Avery, Capt. L. S. Curtis, E. Crompton, R. Glentworth, Mrs. O. Godwin, G. V. Jacks, R. L. Lucas, H. C. Moss, P. H. Nye, D. A. Osmond, P. Selwood, R. T. Shepherd, P. J. Watson, and R. Webster.

The soil map overlays and air photographs are presented by the courtesy of Hunting Aerosurveys Ltd.

Finally, I am indebted to those numerous friends at home and abroad who, when on leave or when occasion allowed, used to 'drop in' to discuss their work and in consequence gave me so much valuable information.

G.R.C.

Contents

List of Plates

FRONTISPIECE: This little figure in coloured clays was presented to G. R. Clarke by his colleagues on his retirement in 1962 after fifty-two years in soil science. It represents him at work in a soil profile pit. The piece is the work of Audrey Blackman and is inscribed: *The Study of the Soil 1909–1962*. (Photograph by Michael Shiel.)

Introduction

THE study of soil in the field comprises the detailed and accurate observation of the soil as a whole in relation to the various forms of life with which it is associated. The study in the laboratory of the material obtained by the sampling of a soil pit is an essential adjunct to soil study, but it is not the study of the 'living soil'. So soon as a soil sample is removed from its environment it 'dies', and though laboratory investigation is necessary in order to obtain information not otherwise available it is essentially of the nature of a post-mortem examination. Soil scientists should realize that the story of a soil is told by an examination of it in its natural environment. The observations of the land worker have been used to guide the utilization of the soil from the earliest ages, but these very observations that have meant so much to him have too frequently been neglected or discounted by the laboratory investigators of the last half-century. The idea that a simple chemical analysis of a 'dead' soil sample taken in an arbitrary manner would solve the problems of economical fertilizing and cultivation is now fortunately nearly extinct. In its place is growing the concept of the soil as a live substance, almost an entity.

A soil grows, develops, and responds to environment, care, and sympathetic treatment much as an organism does. History is full of examples of man and soil evolving together. Like other organisms, a soil possesses special characteristics by means of which it can be recognized and classified.

The *form* of a soil is recorded as its *soil profile*. The soil profile as seen on the face of a pit or cutting is the manifestation of all the changes, growth, and development that have taken place during the 'life' of the soil. It may be studied by any intelligent observer when he works on the land or digs a hole in the earth. The pedologist must be a naturalist in sympathy with his subject. He must know what to look for, and how to interpret the true significance of his observations and utilize them for his study.

Until a subject of study is classified, knowledge of that subject is incomplete, so the first and fundamental duty of pedologists is the classification of soils. Many efforts have been made to standardize a system of soil classification, nearly all of the more recent of which have been tolerably successful in the countries of their origin. Soil utilization schemes and maps have resulted. They differ in their systematism and nomenclature, but they all recognize the necessity for the profile pit as the picture of the life history of the soil. For the rational utilization of land, to which end all pedological researches should ultimately be directed, a soil-profile survey is essential.

The ideal to which the pedologist must strive is the study of the soil profile in relation to its environment. The soil environment includes the natural or cultivated vegetation since, until the weathering complex of the mineral material has born vegetation and obtained its complement of humus and become 'live', the material is not soil. 'Humus' comprises not only the products of the decomposition of vegetable matter but also the living organisms that bring about the biochemical reactions of this decomposition. Newly weathered rocks, desert sands, or sea sands, until some form of life is established, for example fungi, bacteria, etc., or until they become anchored by vegetation and humus, cannot be soils or develop into soils. Vegetation and soil development cannot be dissociated; they are, in fact, completely merged into a natural living unit.

Dokuchaiev (1879) gave the first definition of 'soil' as we now understand it:

Whether we admit that the south-western portion of Russia was submerged under the sea in the beginning of the post-tertiary period, as some geologists think, or it was covered by glaciers, as other geologists think, or it was dry land, as still another group of geologists think, matters little. For us it is important that after this or the other of the given phenomena the upper layers of the soils were apparently subject to various processes due to weathering and to processes due to vegetation; both of these were instrumental in changing the upper horizon of the parent material to a greater or less depth. These parent materials which have undergone changes by the mutual activities of air, water and plants, I call soil.

1. Soil Site Characteristics

MAN, in his endeavour to use the soil for his needs, was compelled to adopt certain methods for the recognition of what he required. His choice was not always strictly according to the rules of natural site selection, but from the place-names of old settlements or from archaeological excavations one is compelled to believe that early man had a very shrewd idea of the best land units for his simple needs, for example well drained soils, not acid, and with a clear water supply.

A unit of land suitable for a single uniform system of utilization may be termed a 'site' (Bourne 1931). Man has always recognized 'site' but has not always realized how he has arrived at his conclusions. In the appraisal of a soil for his needs, man consciously or otherwise assesses its value from a consideration of the same general characteristics as were observed by Dokuchaiev towards the end of the last century during his attempted soil utilization survey of Russia (Afanasiev 1927). He observed that every 'dry land vegetative soil' is a result of the functioning of the following factors:

the climate of the locality,
the nature of the parent material,
the mass and character of the vegetation,
the age of the landscape,
the relief of the locality.

He concluded that when all the above factors were similar in two different localities, no matter how far apart, the soils would be similar. That is, they have a similar life history and are pedogenetically related; or, in still another way, if environmental conditions in two different localities are similar the soils are similar. These conclusions lead on to the natural corollary that if all the soil-forming factors or site characteristics are known, then the soil-profile characteristics may be foretold. Certain other factors have also to be considered, especially man's influence and management of natural sites in the past or for the immediate or distant future. Assuming that Dokuchaiev's factors are the minimum possible number of observations for the assessment of the soil site, it is necessary briefly to discuss the part played by each factor in the production of the complete whole.

TABLE 1

A suggested proforma for the recording and indexing of the characteristics of a soil site

Index data

KEY REF.:   *Surveyor:*   *Date:*

Genetic group: *Soil category:*
Soil catena name: *Catenary unit:*
Soil series or association: *Type:*   *Phase:*
Air cover refs.:

Soil site characteristics

I.	*Locality:*	map ref.:	map symbol:

II. *Age of landscape:*
III. *Topographic data:*   altitude, slope, and form
                              aspect, exposure, and micro-relief
IV. *Drainage:*
V. *Parent material:*   stratigraphic age, lithological nature, geological bed, mode of formation and origin of superficial deposits
VI. *Climatic data:*   major type, rainfall and distribution
                              temperature and distribution
                              frost periods and prevailing winds
VII. *Vegetation:*   plant formation ⎫ natural species
                              plant association ⎬ semi-natural species
                              plant communities ⎭ cultivated species
VIII. *Fauna:*   natural species ⎫ density of population and distribution
                              semi-natural species ⎬ tion and distribu-
                              cultivated species ⎭ tion
IX. *History:*   special reference to human influence past and present
X. *Deleterious factors:*
XI. *Scaled sketch of relief profile*

The reverse side of the form deals with the soil profile description (Table 2).

Table 1 illustrates a proforma on which the soil site may be described and recorded, while Table 2 (Chapter 2) provides one for the study and recording of the soil profile. It is not suggested that they represent a perfect schedule of observations but the general system, with modifications to suit particular environmental conditions, is applicable throughout the world.

THE SOIL SITE DESCRIPTION

Index data

References on the top line of Table 1 are so obvious that no explanation is needed. It is equally obvious that the soil and site classes cannot be entered up until all the soil and site characteristics have been recorded, analysed, and classified. The names used for the genetic groups and categories of the soils should be those in general use. The others are more closely associated with local environmental factors, and their precise nomenclature depends upon the system of classification used locally.

Air cover references

When air photographs are used it is desirable that their particulars should be recorded for future reference. Air cover references should include the details specified in Chapter 5 (pp. 104–5).

SOIL SITE CHARACTERISTICS

Item I. Locality of region, catena, and site.

Each soil profile is representative of a certain soil site, and the association of a sequence or pattern of sites constitutes a *soil region*. The soil region is discussed in great detail by Bourne (1931). In a country for which adequate maps are available the boundaries of a soil region can be delineated comparatively easily. The main watersheds and catchment areas stand out as simple, natural, topographical regions and, if a place-name be given in conjunction with the latitude and longitude, the locality is clearly defined.

The term *catena* has been used by Milne (1935) to express a sequence of soil profiles that appear in a regular repetition in a region possessing a regular succession of certain topographical features. A common example of a catena occurs in the regularly undulating lands of the tropics, where the soils on the ridges may be

2

red loams, with murram (iron concretion) soils on the upper slopes
and black soils in the bottoms, or valleys.

Fig. 1 represents such a sequence recorded by Milne during a
reconnaissance survey in Tanzania (1947). The accompanying
text explains the general principles of his use of the term.

At Ukiriguru (and the pattern is repeated throughout its neighbourhood) the soils provide a good example of the catena or

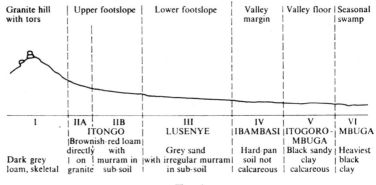

FIG. 1

topographic sequence of types, the zones running in succession
from the flanks of the rocky hills to the lowest parts of the intervening
valleys. The sequence is as follows.

I. Hill-tops and rocky parts of slopes: scanty skeletal grey loam in
crevices of rocks. Vegetation probably originally *miombo*, for a specimen
of *Isoberlinia globiflora* was seen, but there is nothing now but a light
growth of secondary bush.

II. Zone at foot of rocks, 100–300 m wide across the line of greatest
slope, but may be locally more extensive, as on slopes of larger hills or
saddles between them. Coarsely gritty brown to red-brown loam *itongo*
soil of variable depth, based directly upon the granite in the upper part
of the zone but having a hard clinker-like *murram* horizon at 1·5 m depth
lower down the slope. The whole profile is slightly acid. Only remnants of
the woody vegetation remain, including *Afzelia quanzensis, Tamarindus
indica, Dalbergia melanoxylon, Kigelia*, and *Terminalia* and *Commiphora*
species.

III. Next zone below, where the slope is perhaps 1 in 50: very sandy
pale grey *lusenye* soil, acid in reaction, of variable depth, sometimes based
on the underlying bouldery granite, sometimes with a coarse grey-black-
yellow sandy *murram* as an intervening horizon. The one feature or the

other seems to depend on whether the underlying rock is sufficiently rotted to be porous, or is impervious. The grey colour is due to the high wet-season water-table. Natural vegetation not seen, chief crop in cultivation is cassava.

IV. Next zone below, approaching the floor of the valley: dark brownish-grey gritty clay with 'hard-pan' profile, termed *itogoro-ibambasi*. Seen after heavy rain, this soil was saturated in the top 40 cm, but quite dry below; water seeped out copiously at the depth named when a pit was dug into the dry subsoil. At 1 m depth the subsoil is a dry gritty grey-yellow clay, not calcareous; the whole profile is acid, much more so in the top 60 cm than below. Original vegetation not seen, cultivated to grain crops.

V. On the valley floor, but not quite the lowest part: heavy inky-black sandy clay, wet to the full depth of a 1·5 m pit at the time of sampling, and showing no hard-pan horizon. This clay does not seem to crack much when dry; seen wet, there were no visible structure lines. Calcium carbonate occurs at all depths, in the form of particles that are so numerous from 120 cm downwards that they give a whitish cast to the colour of the subsoil. There is no definite nodular horizon, but the white particles include some large concretions below 120 cm. This soil is given the name *itogoro-mbuga*. Ordinarily such land is grazed only, or planted to sorghum.

VI. In swampy reaches of the valleys there are local areas of the heaviest black cracking clay, known as *mbuga ya bugado*, carrying stands of *Acacia seyal*.

Such a sequence of soils would appear over and over again in consecutive valleys and the profiles encountered in parallel with the drainage ways would be uniform. A similar succession over much shorter distances occurs in the south-eastern steppes of Russia, where the 'motley' soils described by Kostychev occur. In this region 5 m strips of chernozem alternate with similar strips of saline soils. The natural vegetation of the chernozem is the typical steppe feather grass (*Stipa stenophylla* and *S. capillata*) whereas the grey (salty) steppe is associated with *Artemisia maritima*. When the whole sequence is ploughed and put to wheat, the crop grows to a height of 1 m and is of good quality on the chernozem, whereas on the grey saline soils it only grows to about 25 cm and then turns yellow and withers.[1]

The catena is a very valuable addition to our soil vocabulary and provides a useful unit in reconnaissance soil mapping. Recently the catena has been used to embrace certain other combinations of

[1] In such situations micro-relief is of great importance (Sokolovsky 1933).

soils that would not originally have been so described by Milne, as in areas where numerous outcrops of geological beds of different lithological character occur in rapid succession down the slope of some major feature. In such conditions a *mixed catena* (Clarke 1954) may develop, as on the Jurassic rocks in Oxfordshire. The Northampton Sand lies on a hill-top over the Upper Lias Clay. This overlies the Middle Lias Marlstone which in turn overlies the Lower Lias Clay outcropping at the bottom of the hill. The soils in this sequence comprise

(1) Northampton Sand *in situ* with excessive drainage;
(2) Northampton Sand downwash on to Upper Lias Clay with
     (a) a deep well-drained phase and
     (b) a shallow ill-drained phase;
(3) Upper Lias Clay *in situ* with imperfect drainage;
(4) Upper Lias Clay downwash on to Middle Lias Marlstone with
     (a) a deep ill-drained phase and
     (b) a shallow well-drained phase
(5) Middle Lias Marlstone *in situ* with free drainage;
(6) Middle Lias Marlstone downwash on to Lower Lias Clay with
     (a) a deep well-drained phase and
     (b) a shallow ill-drained phase;
(7) Lower Lias Clay *in situ* with impeded drainage.

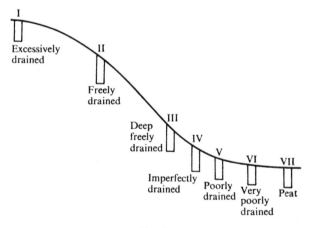

FIG. 2

The soil association used by the Soil Survey of Scotland (Glentworth 1954) is a practical development of the catenary concept. The basic unit is the hydrological sequence (Fig. 2) over a relatively uniform parent material in one climatic zone. This is subdivided into soil series according to drainage. Thus it is a true catenary classification in the sense that Milne originally envisaged it.

Item II. Age of site

This item is of importance as it affects the degree of maturity of the soil. It is the soil in equilibrium with its present environment that must be described, and all archaic influences, unless still producing effects, must be discounted. For example, it would be incorrect to describe a soil as a Brown Forest Soil if the original deciduous forest had been cut and cleared so long ago as to have allowed the soil to become degraded in all the pedological attributes that it possessed when in equilibrium with the forest. This despite the fact that it would revert to forest and become regraded in quite a short time if left to natural influences.

It is desirable to determine whether the soil is the result of old and practically stable conditions (residual soils) or is the result of some more recent geological, pedological, or other process. The following hypothetical chain of events in soil evolution is somewhat improbable, because of side issues and the probable influence of other factors, but is made to emphasize the importance of the time factor.

Imagine a vast tract of new land of variable topography, with a variety of geological formations outcropping, under a cool humid climate with an evenly distributed rainfall. If the rocks are at all permeable and are weathered to a detritus on which some vegetative growth is possible, soils will develop in course of time. Under these climatic conditions, the percolation of rainwater will tend to exceed the surface evaporation. A steady washing-out naturally ensues of the finer mineral particles from the surface horizon, and also of those materials soluble in water, carbonic acid, and organic substances produced by the decomposition of vegetable residues. A time will come, therefore, when the surface material or soil has become almost destitute of all mineral material except quartz, while the underlying material may become proportionally richer in some of these substances that are thrown out of solution as insoluble substances which hence accumulate. On fairly flat topography this precipitation and accumulation may continue until further percolation is impossible, and either the water must escape by lateral drainage or bog conditions will prevail. (Such cases do in fact occur on Dartmoor, where a secondary 'swamp soil' is built up upon the accumulation zone of an archaic podzol). During the evolution of the soil each stage in its formation is accompanied by a specific vegetational climax, some of which tend to prevent, while others tend to foster, soil movement or 'creep'. The end of the one cycle comes when the loose material of the surface horizon of the upland site has slipped down into the valley and left the accumulation zone exposed to further aerobic weathering. (Such truncated soils are quite common in various parts of the world.) Another vegetational climax becomes established and weathering proceeds, while wind, erosion, and creep continue to remove the loose unanchored products towards the lower sites. (The outwash of soil from uprooted trees and from workings of ground fauna is quite appreciable on hill-side sites.) Fig. 3 illustrates these hypotheses. Fig. 4 represents an actual land-form described by Topham (1937) in Malawi.

Thus by the cumulative effects of environmental and soil-forming processes, the topography of the land becomes changed with the elimination of irregularities and the formation of smooth curves. Such changes in gross topography may be followed by changes in climate, and vegetational climaxes.

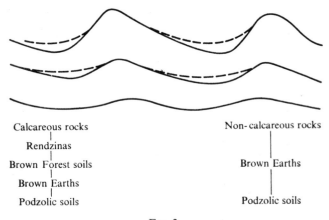

Calcareous rocks	Non-calcareous rocks
Rendzinas	
Brown Forest soils	Brown Earths
Brown Earths	
Podzolic soils	Podzolic soils

Fig. 3

Item III. Topographic data: altitude, slope and form, aspect and exposure, surface relief

It is difficult to analyse the relative importance of the various soil-forming factors, but undoubtedly topography is frequently the dominating influence controlling the local climate, vegetation, and regional drainage.

The general lie of the land or *main* topography strikes the observer directly he enters a region, and such outstanding features as the vertical zoning of vegetation on mountain-sides, or flat plains with

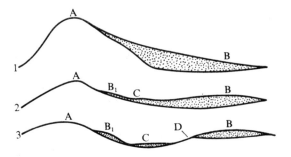

FIG. 4 A common landscape in Malawi

The three diagrams represent stages in the weathering of a hill A. The shaded portion B is a quartz drift derived from A; B_1 is a small secondary drift. The main mass of B becomes separated from A by erosion of the drainage basin C.

At the present day, as depicted in the third diagram, B is leached sandy hill-top; at D there is an outcrop of *lubwe*, concretionary ironstone; B_1 is a drift soil but less leached than B; it is often fertile, normal erosion being rapid. B is infertile and carries *Uapaca* forest.

meandering streams, obviously lend themselves to the formation of characteristic soil sites. In general it is enough to classify the main topography into

Mountains	*Upland plateaux*	*Main escarpments*	*Plains*
Peak	Peak	Hillsides	Ridges
Plateau	Plateau	Crest sites	Isolated hills in
Mountain- side	Valley bottom	Escarpment shelves	plains Flat plains
Valley bottom	Peneplain		Easy hill-sides Basin sites Riverain sites

The character of soil sites within such groups is determined partly by aspect. For instance, the soils on the northern and southern slopes of the southern Alps differ very greatly, both in colour and profile. On the great loessial plateaux of China, which have been deeply dissected by rivers, the topography consists of steep hills with flat tops between deep and narrow valleys. The hill-sides facing south are subjected to intense drying-out after the rains, whereas the northern slopes are cooler and more humid and allow of a luxuriant growth of vegetation. Because of this the soils on the north hill-sides are dark and humic, while those on the south slopes are chestnut-coloured and less fertile. Again, the climate towards the summit of even a 1000 m mountain varies sufficiently from that in the valley to produce a quite different natural vegetation and soil type. In the soil description, therefore, the first remarks on topography must classify the soil site according to altitude.

Such groups, however, are generally capable of a further division on what may be termed *mezzo-relief*. Mezzo-relief includes those smaller topographic features that come to the notice of the observer in a second glance, such as the gentle undulations of a plateau, the slope of a peneplain, or the dip and escarpment slopes of small geological outcrops. The soils formed on such sites will depend to a greater extent upon slope than upon variations in altitude and aspect, though the latter may exert some influence in affording shelter against prevailing winds.

The general lie of the land both for main and mezzo topography may be obtained from the contour lines of the base map, or form lines may be drawn on aerial photographs.

Finally, the pedologically important *micro-relief* must be taken into consideration. Micro-relief includes those small changes in the relief of an otherwise even surface which are more often than not only noticeable after careful inspection. It is more intimately connected with local drainage than with either the altitude, aspect, or slope of the main topography. The soil types resulting from differences in micro-relief are frequently only those of the wet phase or the dry phase of the main soil type of the region. In semi-arid or arid areas and in waterlogged or tidal areas, however, a few centimetres in relief may be directly responsible for an entirely different soil type and its vegetation. Sometimes the features of the micro-relief may be the *result* of processes within the soil mass and may not necessarily be the *cause* of them. In certain countries where heavy montmorillonite clays are found in association with a high soluble salt content and a great capacity for swelling on wetting, the lateral pressure engendered forces the soil mass to buckle into peculiar hollows and hummocks. Some of these soils have been the subject of study in Australia, where they have been given the name of *gilgai*. Micro-relief features have been recorded from arctic regions where the pressures produced by ice lenses and wedges give the polygonal soils of Alaska, etc.

Altitude

Where good topographical maps are available altitude can be deduced from a study of contour lines and spot heights. In the absence of a contoured map altitude may be determined by the field aneroid.

Slope and form

Slope may be defined as the gradient between two points of elevation and may be deduced from the contour lines of a map or may be read in the field with an Abney level. Slope may be measured and classified in various ways but the usual method is to record it either as a percentage slope or degree slope in the *field*, or as a gradient from a *map*. The percentage slope is the number of units rise or fall in 100 units of horizontal equivalent (not the hypotenuse). A 45° slope is equivalent to a 100 per cent slope because it is 100 units in 100 units, or 1 : 1 as a gradient. There are various other field descriptions for slope depending upon the use to which the site may have been put, but for soil surveyors it is better to record the actual measurements.

It is frequently the case that the *form* of the slope is more important to the soil surveyor than is the overall gradient. Form may first be divided into two main classes, viz. (a) uniform or even slopes and (b) complex slopes. The latter can further be divided into

terrace sequences,
undulations in one direction,
undulations in various directions,
concave,
convex,
concavo-convex and all combinations, right down to the 'broken topography' of landslides, boulder moors, etc.

When describing complex slopes the uppermost portion should be named first, for example plano-concave, where plane is at the top and concave lower down. When mapping soils in areas over which contour lines are few or are at wide vertical intervals it is often useful to record on the field map certain symbols that indicate surface characteristics. Such symbols will never reach the final map but can assist the surveyor in writing up the memoir.

```
Flat ──── or left blank.
Undulating  ∿∿∿∿
Broken and irregular  ∿w∿w
Terraces, lynchets, &c.  ⌐L

Slopes  0 – 3°    ──────────→  ⎫
        3 – 6°    +─────────→  ⎪
        6 – 18°   ++────────→  ⎬  Length of arrow covers extent
       18 – 25°   +++───────→  ⎪  of slope with the head downhill
      over 25°    ++++──────→  ⎭
```

A scale diagram representing the site in relation to its topographic environment should always accompany the written site description.

The Soil Survey of Canada recognizes six main classes of relief. This system has also been adopted by the Soil Survey of Scotland (Glentworth 1954). The classes are as follows:

Site areas:
(1) depressional to flat (chiefly 0°–0·3° slopes),
(2) very gently sloping to gently undulating (0·3°–3° slopes of low frequency),
(3) undulating (1°–3° slopes of high frequency),
(4) gently to moderately rolling (3°–9° slopes),
(5) mixed undulating and rolling,
(6) strongly rolling to hilly (9°–17° slopes).

Such slope data would normally be obtained in the field by means of an Abney level. When well-contoured maps are available a good deal of information may be obtained by direct interpolation; much of the mezzo- and probably all of the micro-relief may be intra- or sub-contour.

Aspect and exposure

Aspect is the direction of a compass bearing at right angles to the feature and will be sufficiently accurate for most practical purposes if taken to the nearest sixteenth of the circle. It may be expressed either in degrees or cardinal points. Aspect infers some degree of slope and it is very difficult to determine the lowest degree of slope that can still be said to retain an effective aspect. In site evaluation, however, it may be assumed that slopes of less than 3° possess 'no aspect'.

Shelter or shade due to relief should be noted as it may exert a deleterious effect upon the local micro-climate. For example, at Edinburgh the angle of the sun's elevation on 26 December is 12°, on 24 January it is 15°; this means that the north side of a moderately steep hill 100–150 ft high, 350–500 ft O.D. is sunless for 3 months of the year (November to February). Exposure, too, may exert similar effects, particularly with respect to winds and storms. In fact the size, shape, and nature of the topographical features of a site and its surroundings may well be the controlling factors in its specific phenology. The colonization by ash of the northern, i.e. cooler and damper, slopes of some of the calcareous hills of England, while the southern, warmer, and drier slopes do not carry ash is a good example of this phenomenon. J. M. B. Brown (1953) gives many interesting details relating to the importance of site characteristics upon some of these phenological problems.

Item IV. Drainage of site

Site drainage is concerned with the removal from the soil of all that water, from direct precipitation or from other sources, that is not evaporated, transpired, or removed by run-off. This drainage is dependent upon the disposition of the watersheds and drainage basins of the particular geomorphological region, and on the permeability of the soil and strata. Note should be taken of the liability of the area to seasonal flooding or to tidal inundation, and of the possibilities of sheet or gully erosion, evidence of which should be looked for in outwash fans of soil material on the lower-lying sites. Any artificial methods of protection or prevention should be described under Item IX (History). The frequency of the streams and whether they are free or 'locked' (i.e. whether they are controlled by locks in such a manner as to exert any effect upon the water-table) is always of importance. The height of the permanent water-table should be ascertained.

Some confusion occasionally arises in the use of such terms as *ground-water, water-table*, etc., and to clarify the situation the following paragraphs and figure are quoted directly from Holmes (1944).

The following scheme shows the various ways in which rainwater is distributed.

Run-off	Direct flow down surface slopes	
	Superficial flow through soil and subsoil to streams	Stream flow
Percolation	Downward infiltration into bed-rocks to replenish the ground-water and maintain its circulation	Ground-water
	Absorption by soil and vegetation subsequently evaporated	Evaporation
Direct evaporation		

(During weathering a relatively trifling amount of water is fixed by hydration in clay minerals and other weathering products.)

Ground-water supplied by rain or snow or by infiltration from rivers and lakes is described as *meteoric*. Fresh or salt water entrapped in sediments during their deposition is distinguished as *connate*. During burial and compaction of the sediments, much of this fossil water is expelled, and during metamorphism most of it is driven out, carrying with it dissolved material which helps to cement the sediment at higher levels. Steam and hot mineral-laden water liberated during igneous activity, and believed to reach the surface for the first time, is known as *juvenile* water.

The storage and circulation of ground-water

Below a certain level, never far down in humid regions, all porous and fissured rocks are completely saturated with water. The upper surface of this ground-water is called the *water-table*. The water-table is arched up under hills, roughly following the relief of the ground, but with a more subdued surface. In general, three successive zones may conveniently be recognized (Fig. 5):

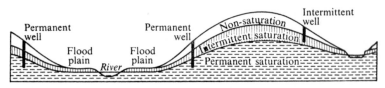

Fig. 5 Underground Waters

(a) *The zone of non-saturation*, which is never completely filled but through which the water percolates on its way to the underlying zones. A certain amount of water is retained by the soil, which yields it up to plant roots.

(b) *The zone of intermittent saturation*, which extends from the highest level reached by ground-water after a period of prolonged wet weather, down to the lowest level to which the water-table recedes after drought.

(c) *The zone of permanent saturation,* which extends downwards to the limit beneath which ground-water is not encountered. The depths in mines and borings at which the rocks are found to be dry vary very considerably according to the local structures, but a limit of the order 650–1000 m is not uncommon. Juvenile and expelled connate water may, of course, ascend from much greater depths.

Wherever the zone of permanent saturation rises above ground level, seepages, swamps, lakes, or rivers occur. When the zone of intermittent saturation temporarily reaches the surface, floods develop and intermittent springs appear. Conversely, many springs and swamps, and even the rivers of some regions, go more or less dry after long periods of dry weather when the water-table falls below its usual level.

Superficial drainage is mainly a function of the mezzo-relief, and is particularly important in the case of heavy clays on slight slopes where the material is so impermeable that the surface run-off exceeds the direct percolation. Such soils, even in a cool humid climate, frequently possess the characteristics of arid or semi-arid types in that they exhibit in their lower horizons concretionary or secondary calcium carbonate and sulphate.

The following terms may be used to describe site drainage.

Satisfactory. Water from any source is without any deleterious effects.
Seasonal drought. Water run-off and evaporation exceed supply. (Deleterious.)
Seasonal wetness. Water from some source remains on or near the surface for appreciable periods of time. (Deleterious.)
Permanent wetness. This is usually associated with basin or valley bottom sites and infers some physical impedance to the free flow of water.
Liable to flood. This is usually connected with seasonal wetness but is frequently associated with riverine or lacustrine sites.
Spring lines, flushes, seepages, etc.
Irrigated.
Warp.

Sites may further be classified on the ease or otherwise with which the drainage could be improved artificially. For example, a seasonally wet site might easily be improved by a local drainage scheme or, on the contrary, nothing but some major regional scheme such as the dredging of a river, would suffice. The speed with which water removes itself from the site is also of significance. When water on a slope is in excess of that which will percolate into the soil it may sometimes run away without injury to the soil, but if the slope and the nature of the soil allow, erosion will ensue. This contingency should not be overlooked anywhere. In landscapes liable to erosion elaborate systems of classification are called for; it will suffice here to deal only with the simplest. Use may be made of symbols on the field sheet for four categories.

SR +++ Surface run-off intense—erosion probable.
SR ++ Surface run-off definite but serious erosion improbable.
SR + Surface run-off is purely local and no abrasives are carried from elsewhere, but local 'carry' may silt up drains and ditches.
SR 0 No run-off detectable.

Item V. Parent material group: stratigraphy, lithology, origin of superficial deposits

The field-man often has difficulty in determining whether or not the soil material with which he is working is truly derived from the rock immediately beneath. The parent material (P.M.) is the unconsolidated mineral mass from which the soil itself has derived and is referred to as the *C horizon*. It is the material from which by pedological processes alone the *in situ* soil profile has been developed. It need not bear any relation to the lithological character of the geological beds that may be found immediately beneath it, but if it does then the soil is described as having developed *in situ* and is called a sedentary or 'straight' soil. In countries such as England, where most soils may be said to be in a state of stable immaturity, the nature of the mineral material from which the soil is developing is of great significance and is often the basis for its classification as an intrazonal soil. This class can then be further subdivided into soil series specifically related to local geological outcrops.

There are numerous systems for the classification of the rocks that give rise to parent materials. However, the 1936 list is still useful with the addition of a class for superficial deposits.

In addition to these groups allowance is also made for variation of material within them. For example, in group 1 of acid igneous rocks, there would be subgroups for the various acid igneous rocks such as granite, quartzose schist, etc., or drift containing acid igneous rock and hard shale together. Similarly 'sandstone, non-calcareous' would include red or yellow sandstones and their drifts.

List of parent rock materials

Group no.
1 Acid igneous rocks
2 Basic igneous rocks
3 Ultra-basic igneous rocks
4 Basic tuffaceous shale
5 Schist and gneiss
6 Slate and hard shale
7 Clay and clay shales, calcareous
8 Clay and clay shales, non-calcareous
9 Red clay, calcareous
10 Red clay, non-calcareous
11 Silt-clay
12 Sand, calcareous
13 Sand, non-calcareous
14 Sandstone, calcareous
15 Sandstone, non-calcareous
16 Sandstone, felspathic, non-calcareous
17 Glauconitic sand
18 Hard limestone
19 Soft limestone
20 Chalk
21 Brickearth
22 Peat
23 Complex drifts: e.g. clay with flint
24 Alluvium

$$\text{Old alluvium} \begin{cases} \text{calcareous} \\ \text{non-calcareous} \end{cases}$$

$$\text{Recent or active alluvium} \begin{cases} \text{calcareous} \\ \text{non-calcareous.} \end{cases}$$

Recent alluvium may be further subdivided on its activity:
(a) active alluvium—still accumulating mineral matter,
(b) intermittently active alluvium,
(c) stable alluvium—no longer receiving additions of mineral matter.
It should also be described as of freshwater or brackish origin.

Peat soils are subdivided into

$$\text{Basin peat} \begin{cases} \text{Basic—B} \\ \text{Acid—A} \end{cases}$$

$$\text{Moss peat} \begin{cases} \text{Blanket moss—C} \\ \text{Hill peat—H} \\ \text{Raised moss—M} \end{cases}$$

Raised moss develops either on other peat or on other soil types,

e.g. Raised moss peat developing on acid or basic basin peat; $\dfrac{M}{A}$ or $\dfrac{M}{B}$.

As the mapping of Britain proceeds it is becoming a more frequent practice to record not only the P.M. group but also the proper name of the geological series or bed on which the soil site has developed. This is a particularly valuable addition to the site record since one P.M. group, for example No. 19; soft limestones, can give different soil series on almost every kind of soft limestone encountered. For example, oolitic limestones of the Jurassic system such as the Inferior Oolite limestone (G_5), the Cornbrash (G_9), and the Coral Rag of the Corallian limestone (G_{11}) act as parent materials for the Sherborne, Blenheim, and Marcham soil series respectively.

Sites may also be described as of composite P.M. profile, such as thin washes of one material over material of a different group or subgroup, but such composite P.M.s must not be confused with the *pedogenic* profiles.

The presence or absence of drift material is of great importance since, although the drift may not be derived from the rock beneath it, yet it is obviously the parent material of the soil. Deep drift or old drift can act as parent material in soil formation as readily as a true geological rock.

In field work there are, however, many aids to help the surveyor to gather a rough idea of the source of his material. The geological map is of great assistance, but should not be too closely followed without giving full regard to the recent topography. The older types of geological map usually ignored just those depths of surface material in which man is most interested. Large areas of drift may usually be recognized by their differences in colour and bedding, and in the chemical and physical nature of the transported stones and minerals. For instance, when a quantity of quartzose pebbles and sand is found mingled with the surface layers of sedentary soil on deep sedimentary clay, it may be assumed that the soil should be described as 'sedentary clay modified by quartz drift'. The modifying material may be of sufficient depth to become the dominating factor in soil formation, in which case it would then be described as 'deep quartzose (gravel) drift overlying sedentary clay'. Greater difficulty usually occurs, however, in recognizing and describing 'local' drift. Local drift usually occurs as thin smears of differently constituted mineral mixtures distributed with varying degrees of intensity all over the topographical region, and may usually be recognized in the features of the mezzo-relief. The importance of local drift is well shown in the mixed catenary sequence discussed on p. 8.

Mode of origin of superficial deposits

This section refers mainly to superficial deposits which are products of weathering during and after the Glacial epoch, for example, lacustrine clays, wind-blown sands, etc.

The surveyor must endeavour to determine from the general lie of the land and the nature of the geological material whether the soil is formed on a sedentary or massive rock weathered *in situ*, or whether it is on material of more recent origin. There are four main agents that are responsible for the laying down of the 'recent' parent material, as distinct from *in situ* weathering:

(1) Ice gives glacial drift (till or boulder clay, moraines), and fluvio-glacial deposits (outwashes of gravels, etc.).
(2) Water gives alluvium, terraces, and footslope deposits.
(3) Gravity gives colluvium (by creep or slump).
(4) Wind gives brickearth, loess, sand drifts, etc.

The clue to which of these groups the P.M. owes its formation is frequently to be found in the size, shape, and colour of the sand grains or the stones, taken in conjunction with the topography and geological nature of the underlying rock.

Wind-borne loess is commonly unstratified, and sand may be fine and rounded; quite frequently they occur in deep folds where wind force was reduced by obstructions, for example on the lee side of topographical features.

Certain other characteristics are frequently detectable, such as the faceting of pebbles by sand blast, the matt surface of sand grains, and the absence of mica flakes. Loess anchors in the vegetation of the steppes. Successive generations of grass grow through this and old root-channels frequently become filled with secondary calcium carbonate while the horizontal banding resembles buried soil horizons.

Colluvial soils usually show themselves as an average of the components of the surrounding sites; stones and sand grains may be angular, or round but, unless they are rounded in the original rock mass, they do not usually travel far enough to undergo much change in taking up their new position.

3

Water. Water deposits are probably by far the most numerous and variable, and it is fairly safe to assume that, after the elimination of wind and glaciation as possible factors, the dominance of rounded and rubbed sand grains signify a water deposit.

Ice. The effects of glacial or fluvio-glacial actions are usually very clearly indicated by topographical features. The lithological character of the pebbles may be distinct from that of the underlying rock. They are usually sub-angular in shape and some bear evidence either of considerable abrasion on one side or, if rounded off, give the impression of having once been rubbed more on one face than the other. They are rarely, if ever, so round as in fluviatile deposits.

Item VI. Climatic data: rainfall, temperature, etc.

The classification and geography of soils on the basis of climate is so widely known that little need be said in explanation of this item. 'Climate' may be divided into main climate and local climate.

The main climate will, to all intents and purposes, be fairly uniform over a large area, and its influence may be accepted as a stable or invariable factor. Main climate is so obvious that in the ordinary selection of a soil site it is taken for granted. Local or micro-climate, on the other hand, is of very great local importance and may be closely related to mezzo-relief. The difference in local climate between the lee side and windward side of a topographic feature may be sufficient to produce differences in the soil profile. Shelter can be responsible for a higher mean temperature or a lower evaporation, and sometimes for the establishment of a different plant association on the lee side from that occurring on the windward side.

Rainfall

In those countries in which the distribution of the annual rainfall is uniform throughout the year a knowledge of its total amount is generally sufficient. In those countries in which there is a marked wet and dry season, however, a knowledge of the monthly distribution is essential, and especially is this so in tropical and sub-tropical regions. No description is complete without these data, and *it is better to state 'unknown' rather than to attempt to surmise a distribution on any geographical grounds.* The form the precipitation takes, whether it falls as snow, hail, or rain, and the period of the year when it is dominant or most intense, are important points to be noted when discussing land utilization.

Temperature

When working over large areas as in the reconnaissance survey of a country the recording of the mean annual temperature, and the mean maximum and mean minimum temperatures with their changes through the year, is about all that can be expected and is probably sufficient. For purposes of detailed soil site evaluation in small and intensively utilized areas many more data are required. The mean monthly variations with the distribution of frost periods and excessively hot periods are the least on which 'local weather' conditions may be assessed. For example, in normal arable areas it appears that plants do not readily grow below a temperature of 5·5°C. Another factor closely related to temperature is the humidity, and any information about this is always worth recording. Frost in certain areas may be a very serious factor and if a site can be labelled as being 'liable to frost damage', or 'liable to valley frost' at certain specified times of the year, it increases the value of the map and memoir for land-users.

Winds

The direction and nature of prevailing winds should be recorded, as they may exert an important influence on soil formation. Exposure of the surface through lack of shelter may allow desiccation and subsequent wind erosion.

In considering the amount of shelter given by various objects it is a useful general practice to assume that the effective limit of a shelter belt on the lee side is usually about ten, but not more than twenty, times its height, and sometimes up to five times the height on the weather side. In the case of sloping objects like hills the effective range is much more difficult to assess since the wind tends to turn back upon itself and by sucking up air from the lee side sometimes produces a vortex effect greater in effect than the prevailing wind. Thus it is possible for a moisture-bearing wind to produce a dehydrating effect upon the lee side of a topographical feature.

If possible, it should be determined whether the maximum wind intensity coincides with the maximum vegetational cover of the soil, or if the maximum intensity occurs at a season of the year when plant growth is at its minimum. The difference in soil and vegetation characteristics resulting from the various distribution periods is fairly obvious.

Item VII. Vegetation

Since the soil and the vegetation that grows upon it are inter-dependent, it is desirable in normal survey work that the soil site selected should so far as possible carry a natural vegetation or, if that is impossible, attempts should be made to ascertain what the vegetation would have been if it had not been changed by human interference.

The vegetational climax (Tansley 1930) and the soil climax, if indeed such do exist, tend to occur and persist for some period of time until the development of, or the invasion by, other vegetation. The age of the site, therefore, will exert some influence on the type and stability of the vegetation, and this naturally becomes reflected in the morphological characteristics of the soil profile. The surveyor should not be too concerned with the definition of the various climaxes, but rather assume, from a study of the age of the trees or the history of the site, that at the moment of observation the natural vegetation and the soil are in equilibrium. In woodland regions the density and nature of the canopy is very important and may be described if 'complete canopy' is expressed as unity and all grades to 'open' or 'no canopy' are expressed as decimals of unity (e.g. 0·5 canopy). If, when the surveyor looks upwards, he cannot see the sky he records 'complete canopy' or '1'. If, on the other hand, he has an open view he records 'open canopy' or '0'. The importance of canopy or cover in the preservation of certain tropical soils is well known. The relationship between rainfall, evaporation, and humidity in effecting characteristic changes in the soil profile is discussed in Item IX.

The density of the population of the various vegetational strata should also be assessed and noted, since the work of Tamm (1920) suggests that the understory and surface vegetation are among the chief agents in organic soil formation.

To assess the density of the vegetation use may be made of the density scales evolved by Burtt (see Swynnerton 1936) during his botanical survey of the tsetse areas. His standards have been tried out in various other localities and for tropical survey in particular have proved to be of great value, both to agriculture and to forest officers. The following table shows how density is assessed.

D_1, impenetrable thicket necessitating the cutting of a path.

D_2, thicket penetrable without the cutting of a path.

D_3, still so thick that a compass traverse is impossible without cutting a path. Grass covering usually about 10 per cent.

D_4, visibility better. Grass covering usually about 20 per cent.

D_5, easy to traverse with compass. Grass covering usually about 30–50 per cent.

D_6, grassland exceeding land covered by bush.

D_7, scattered trees in otherwise open grassland.

D_8, open grassland.

In practice, for categories D_4–D_7 the depth of visibility in yards provides a more suitable description. By the term 'grassland' is meant 'ground not carrying shrubs, trees, or tall herbs'; such ground may or may not actually carry grass; it may in fact be bare, especially in very dry country, or it may be covered by grass in scattered tufts only.

In the description of the vegetation for large area or regional reconnaissance survey the delimitation of the boundaries of the *plant formation* is very useful, and is generally sufficient, particularly in the interpretation of small-scale air photographs. The *plant formation* is the largest unit used in the classification of vegetation and comprises the whole of the natural and semi-natural plant cover occupying a soil region. Many plant formations are quite readily recognizable on air photographs, for example woodland, heathland, fenland, etc.

For site description, however, some smaller and more specific unit is required. In ground survey the recognition of 'dominant' species is sufficient to determine the *plant association* of the site. The *plant association* is the unit next below the plant formation and is usually the most important unit in site classification. This unit embraces the different kinds of plants growing *within* the plant formation, for example *Quercetum roburis* (pedunculate oakwood association). The field description of such a plant association should cover

(1) the species and population of the main standels with approximate estimates of height, girth, age, and canopy;

(2) the species and population of the more abundant plants of the shrub layer;

(3) the species and population of the more abundant plants of the ground vegetation, including mosses.

Similar methods of description should be devised for other plant associations. It must be remembered that the pedologist may, and frequently does, find different soil profiles occurring as small outliers or inliers under isolated *plant communities* and *colonies*. It is, for instance, usually quite easy to detect local pockets of poorly drained soils in the Brown Forest Soil zone by the sharp change in the vegetation from the *Graminetum neutrale* of the former to the *Juncetum* of the latter.

Indicator plants are frequently of use to the pedologist particularly the broader groups such as calcicole and calcifuge plants, saline and alkali plants, 'wet' and 'dry' plants, etc. The following is a short list of indicator plants of the soils of Britain.

ACID HEATH:	*Calluna vulgaris, Erica cinerea, Vaccinium myrtillus*
CHALK:	*Anthyllis vulneraria, Crataegus* spp., *Juniperus communis, Rhamnus catharticus*
LIMESTONE:	*Draba muralis, Sesleria caerulea*
MILLSTONE GRIT:	*Eriophorum vaginatum*
SALT MARSH:	*Glyceria maritima, Limonium vulgare, Salicornia herbacea, Spartina* spp., *Suaedia maritima*
BASE-RICH, POORLY-DRAINED FEN:	*Alnus glutinosa, Calamagrostis* spp., *Phalaris arundinacea, Phragmites communis, Schoenus nigricans*
BASE-POOR, POORLY-DRAINED BOG:	*Drosera* spp., *Eriophorum angustifolium, Molinia caerulea, Myrica gale, Sphagnum* spp.

Digitalis purpurea and *Sarothamnus scoparius* are useful indicators of acid pockets of soil in otherwise base-rich regions.

Among the xerophytic plants indicative of dry areas are *Erodium circutarium, Leontodon taraxacoides, Sedum acre,* and *S. anglicum.* Also found in dry areas are the short-lived annuals *Catapodium rigidum* and *Vulpia bromoides.*

Such indicator species should always be used with some discretion, bearing in mind that it is possible for plants to appear in 'wrong places' due to interference with the base status or water regime of the soil during cultivation.

The speed at which equilibrium between vegetation and soil is attained appears to be much greater than is generally recognized, and the effects of a change of species can be reflected upon the profile in the space of a few years. An interesting example of this occurs in the neighbourhood of Oxford where a series of plots was put down in old forest, comprising some eight varieties of conifers and a number of varieties of broad-leaved trees (Morison and Clarke 1934). The soil is a quartzose gravel over clay, which at the time of planting out was a typical Brown Forest Soil fairly well drained, and with an evenly distributed rainfall of 700 mm. By 1928 all the coniferous sites were slightly podzolized, with each type showing a specific profile, while the broad-leaved sites still retained all the characteristics of the Brown Forest Soil, with a specific A horizon for each variety of tree. Muir (1934) recorded a similar example in the Teindland State Forest.

In addition to the chemical effects of the vegetation upon the soil mass, the physical effects are also important. The expansion by growth of the plant roots—and more particularly is this the case with trees—exerts a definite pressure so that the soil structure and constitution undergo certain modifications. The surface of the soil becomes raised to produce slight differences in micro-relief, and in consequence a modification of the local or micro-drainage. Root pressure tends to alter both the shape and the size of the structural elements, so that lateral cracks increase in number and tend to produce further changes in the constitution by increasing or decreasing porosity and compactness. Root-channels, after the decomposition of the vegetable matter, allow of increased aeration and percolation, so that there is a consequent increase in range of the pedological process.

Item VIII. Fauna: natural, semi-natural, and cultivated

The faunal population of a soil site plays as important a role in the formation of the site as does any other factor. The effects produced by fauna vary greatly according to the particular species involved. Soil fauna may be divided into three groups, depending mainly upon their zone of habitation in, or upon, the soil. The three groups comprise

(1) The intra-soil fauna that include protozoa, worms, insects, etc., which take part in soil-forming processes by living and moving within the soil mass.

(2) The extra-soil fauna, including the birds and all those animals that eat and sleep upon the surface of the earth and produce little or no movement within the soil mass.

(3) Another group comprising those creatures that inhabit a dual environment partially in and partially upon the soil. Such creatures include the burrowing animals such as the rabbit, and numerous insects and their larvae. The effects of these upon the soil mass are frequently of extreme importance.

The faunal population may further be classified in a manner similar to the vegetation, viz.

NATURAL: Truly indigenous, and in permanent equilibrium with the existing soil-forming factors.

SEMI-NATURAL: Originally present through invasion but now established; or established by human interference with the natural environment.

CULTIVATED: Domesticated animals maintained by man, in equilibrium with some system of land use: include man himself during the 'age of cultivation'.

The latter categories may exert such influences upon the former as to change the entire biological cycle and in consequence to change the morphological character of the soil profile. For example, intensive introduction of certain species may alter the balance of the indigenous population by forcing the more mobile species to migrate, and the less mobile to die out altogether. Or again, a site occupied by a healthy population of natural fauna may be looked upon as being in a state of harmony, equilibrium, and stability. But upset one of the factors, for example by the clear-felling and ploughing of broad-leaved woodland on a Brown Forest Soil, and the population of surface-burrowing creatures may be reduced to such an extent by the removal of their food and lodging that the stable Brown Forest Soil will be changed into a metastable Brown Earth.

Cultivated animals such as cattle, sheep, and goats, though fairly ubiquitous in old or highly developed countries, may still be limited to particular sites; for example, hill sheep against lowland and arable sheep.

It is reasonable, therefore, when describing a soil site, to be able to make some constructive, even though tentative, evaluation of the suitability of the site for rational utilization. This is a common practice during soil surveys in new or developing countries (see, for example, C.S.I.R.O. 1960).

The appraisal of damage or liability to damage by vermin or certain species of the natural or semi-natural fauna is obviously a useful addition to the general site description and may justify its inclusion in Item X among the deleterious factors.

The importance of the geographical distribution of certain natural fauna is well exemplified by their effects on the soil-forming processes in the tropics. In such areas termites are probably mainly responsible for the destruction of the surface organic matter. Vageler (1933) estimates that about 100 tons of fresh organic matter per acre falls to the forest floor per annum, and that it does not accumulate is probably due to termite activity. The excretion from such vast numbers of termites brings back to the surface much of the mineral matter from the vegetation they have consumed. The termites' cultivation of fungus gardens in their nests accumulates a rich content of organic matter not to be found in the surrounding unoccupied land. The effect of termite activity in the tropics has been likened to that of the earth-worm in the other climates, but this is not really true. The nature of the material that has passed through the body of the earth-worm is entirely different from the secreted material used by the termites in their termitaria. The life history of the termite is admirably described by Steven Corbet (1935), while many general remarks about their habits and activities are given by Blanck (1931). More recently Nye (1955) and Hesse (1955) have made some important contributions to our knowledge of this subject.

In the cooler climates of the Brown Forest Soils or similar regions, earth-worms, particularly the cast-formers, with moles, mice, or voles, are the chief faunal agents in soil formation. Earth-worms in particular may be regarded as extremely valuable 'indicators'. Worms are, in general, peculiarly sensitive to acidity and are rarely found in conditions of acidity greater than pH 5. Thus much may be learned from their presence or absence. They also act as good indicators of the limits of distribution of humus and air, since they rarely travel lower than is convenient for their needs. For a worm to be found in a true gley horizon is a rare phenomenon and would require some explanation such as a recent lowering of the water-table following a new system of drainage conditions. The value of worms or the numerous burrowing animals as soil-forming factors depends to a great extent upon their aerating and draining capacity, while the bringing of material from below to the surface as casts or mounds improves the constitution and in certain cases may prevent soil degradation by the recalcification of the surface horizon.

The *crotovinas* of the black soil of the great grain belts of Russia, owe their formation to the inwash of surface soil material into the burrows of hibernating animals such as marmots during the melting of the snows.

Item IX. History: human influence, past, present, and future

Man's utilization of natural sites is probably the most erratic and certainly the most rapid of all soil-forming factors. Man and his implements can, in the course of two or three seasons, reclaim a marsh to a meadow or change a Brown Forest Soil to a Brown Earth, while in a slightly longer time he may reclaim and convert a Podzol into an agricultural Brown Earth. Man's influence is usually directed towards one or other of two main objectives, either the alteration of the character of the soil in such a way that a change of the vegetational equilibrium is established, or the changing of the vegetation in such a manner that a change of soil condition will ensue. For the bringing into cultivation of virgin land all sorts of methods are available, but in all cases, except those in which a new soil and vegetation equilibrium is attained, soil degradation must inevitably follow. The shifting cultivation of the tropics or the fallowing of the chernozems are good examples of natural regrading processes, while the necessity for the continuous manuring or artificial fertilizing of land under arable cultivation in order to maintain economic yields of more exacting species of vegetation is a common practice of modern farming. It may be assumed that the natural 'climax' type of vegetation in the humid temperate zone is the broad-leaved woodland, and drastic changes may be wrought in this as a direct result of human influence. First, man may change the type of the woodland formation by establishing a less base-demanding species of tree, with its complement of acidic humus and leaf fall of low base status. The ultimate end of this degradation would lead to the formation of the podzolic soil type. Or, secondly, he may cut and clear the woodland and practise tillage. It is with the second system of utilization that we are more closely concerned because it is more intimately connected with the immediate needs of the human community. The soil components most susceptible to the influence of man are water and organic matter.

It is generally understood that in the broad-leaved forests of the temperate zone the heavy canopy tends to depress surface evaporation, while the trees transpire water removed mainly from the depth. The first result of cutting and clearing broad-leaved forest is to expose the surface of the soil to direct climatic influences and increased insolation and to upset the relation between the upper moist humic zone and the lower wet zone in the soil mass by the increased evaporation and drying-out at the surface. At the same time, the even micro-climate of the forest becomes changed to the rapidly variable micro-climate of open areas. The loss of canopy and of tree transpiration may cause a rise in the level of the ground-water. The exposure and increased insolation also bring about a more rapid decomposition of the organic matter so that the water-holding capacity of the surface layers becomes reduced. The percolating water received by direct precipitation then increases in intensity. Such changes in the water régime alter the distribution of water-soluble substances in the profile, while the chemical and physical properties of the organic matter are also affected. Such changes are generally adverse and are readily discernible in the field in the structural and constitutional features of the soil profile. The cessation of leaf fall of high mineral content allows the quantity of bases in the topsoil to decrease. The loss of organic matter also represses the activities of earth-worms, the presence of which is an important factor in the incorporation of organic matter in the soil mass. That the soil does not quickly degrade to a Podzol is due in the main to two causes. One is that by regular cultivation the top zone of the soil is being continually inverted and the bases that tend to be leached downward a little way are thus brought back to the surface horizon. The second, and probably the most important, is again directly due to man's influence, in adding salts and organic matter to replace that removed from the soil in crops for his needs. The result therefore of cutting, clearing, and cultivating the natural forests of England is to change a stable Brown Forest Soil into a metastable Brown Earth (Clarke 1933). When kept in constant cultivation further changes may occur in these metastable soils by a loss in structure and base status. Base status can only be maintained by artificial management, while the retention of the highly desirable crumb structure of the humus–calcium–clay aggregates is only possible by the addition of humus and lime. Failure to provide these leads to degradation, and ultimately either to infertility or the invasion of an undesirable type of vegetation in equilibrium with these soil conditions.

The maintenance of a good calcium-clay, coupled with a sufficiency of plant nutrients and a crumb structure is the acme of all economic agricultural utilization. Where such cannot be obtained, man must accept the soil he has got and make use of a *formation* of vegetation, such as woodland or grass, which will attain equilibrium and thrive under these adverse conditions. Man may, however, use his soil to a wrong end and, when he does, disaster follows.

Since all soils are 'fertile' for the maintenance of their own natural vegetation it is obvious that the term *soil fertility* implies the rational treatment of soils to produce an increase in yield of the existing vegetation or the development of another kind of vegetation of greater value, while at the same time the natural resources of the soil are conserved. A man-made fertility such as this may be defined as the capacity of a soil to yield adequate amounts of the essential plant nutrients to a growing crop and to allow such cultivation as will create a suitable environment for the growth and development of a crop to an economically useful yield.

Item X. Deleterious factors

In evaluating a soil site for utilization the soil surveyor may come upon certain local environmental factors not specified in his general investigations. The result of the operation of some of these may be sufficient completely to negate the value of all the good points he may have already recorded and assessed. Most of the deleterious factors associated with land utilization not already dealt with in the main questionnaire are directly connected with comparatively recent human influence. There may be almost no limit to the variety of factors that the surveyor may find, but a few may be mentioned, such as

smoke damage,
cement works dust,
chemical factory fumes,
liability to fire,
invasions by people, birds, or animals due to the increase in the urbanization of the locality, etc.

The addition to the record of such phrases as 'no detectable deleterious factors' or 'very serious contamination of vegetation by cement dust' are helpful.

2. The Soil-Profile Pit

THERE is nothing new in the idea of the examination of a soil pit as a means of studying the soil *in situ*. The classic example of the soil pit occurs in Virgil's *Georgic, c.* 40 B.C., which, freely translated, may read: 'First choose out a place and then order a pit to be dug where the ground is solid [true *in situ* material],[1] then throw in all the earth again and tread it well down, if it does not fill the pit the soil is loose [i.e. crumb structure] and will abundantly supply the cattle and fruitful vines. But if it refuses to go into its place again and rises above the pit that has been filled up, the soil is thick [texture and constitution], then expect sluggish clods [structure] and thick ridges and plough up the earth with strong bullocks.' Cheap paper allows the modern observer to record more observations than the older investigators, but there is no evidence to prove that they did not see just as much as any observers could do today. From Virgil's pit to the recorded soil profile takes us through about seventeen hundred years of effort with but little noticeable improvement. Robert Plot, sometime Professor of Chemistry in the University of Oxford, described the soils of Oxford and the profile of an ochre pit on Shotover Hill, near Oxford. In his *Natural history of Oxfordshire*, printed at 'The Theater in Oxford', 1677, he says (p. 52):

As to the qualifications of the Soil in respect of Corn, I find them in goodness to differ much, and not only according to their several compositions (being in some places black, or reddish *each*: in others a clay or chalky ground, some mixt of earth and sand, clay and sand, gravel and clay, etc.) but chiefly according to the *depth* of the *mould or uppermost coat of the earth*, and the *nature* of the *ground next immediately under it*: for let the uppermost mould be never so rich, if it have not some depth, or such a ground just underneath it, *as will permit all superfluous moisture to descend, and admit also the hot and comfortable steams to ascend*, it will not be so fertile as a much leaner soil that enjoys these conditions.

[1] Italics and words in square brackets are the present authors'.

Thus have I often-times seen in the County, in all appearance a very good soil, and such indeed as would otherwise have been really so, less fertile because of its shallowness, and a cold stiff clay, or close freestone next under-neath it, than a much poorer land of some considerable depth, and lying over a sand or gravel, *through which all superfluous moisture might descend*, and not stand, as upon clay or stone, to chill the roots and make the Corn languish.

Where, by the way let it be noted, that I said a cold stiff clay or close free-stone; for if there be under a shallow mould, a clay that's mixed (as 'tis common in the blew ones of this County) either with pyrites aureus, or brass lumps; or the stones be of the warm calcarious kind, it may nevertheless be fruitful in Corn, because these, I suppose, do warm the ground, and give so much strength, that they largely recompence what was wanting in depth.

Then on p. 55 he continues:

They dig it [the ochre] now at Shotover on the east side of the Hill, on the right hand of the way leading from Oxford to Whately, though questionless it may be had in many other parts of it; The vein dips from East to West, and lies from seven to thirty feet in depth, and between two and seven inches thick; enwrapped it is within ten folds of Earth, all which must be past through before they come at it; for the Earth is here, as at most other places, I think I may say of a bulbous nature, several folds of divers colours and *consistencies*, still including one another, not unlike the several coats of a Tulip root, or Onyon.

> The 1. next the turf, is a reddish earth.
> 2. a pale blue clay.
> 3. a yellow sand.
> 4. a white clay.
> 5. an iron stone.
> 6. a white, and sometimes a reddish Maum.
> 7. a green, fat, oily kind of clay.
> 8. a thin iron-coloured rubble.
> 9. a green clay again.
> 10. another iron rubble, almost like Smiths cinders.

During three centuries of side-fall and soil creep this pit of Plot's has completely disappeared.

Preparation of the profile pit

The required site being found and its characteristics recorded, the profile pit may be prepared. It is always better to make a thoroughly good pit while about it, since so often on returning one is apt to wonder if perhaps just another foot deeper might not have yielded still further information. The pit, therefore, should be made to a standard. Though it is often desirable that the surveyor should dig a certain amount in order to familiarize himself with the information to be obtained from the sound and feel of implements during the digging operation, it is obviously a waste of valuable time for a qualified surveyor to be pit digging when he could better be employed in further reconnaissance, sampling, etc.

An important point arises here with reference to the privilege of digging soil pits on private property. Permission to enter and dig is always obtained from the occupier. Then the following procedure was always adopted by the writer both for survey and teaching purposes.

Before the pit is opened two large groundsheets are put down near the side of the hole that is to be 'faced'. If in grassland, well shaped turves are cut to sod-layer depth, lifted out, and placed on the nearer sheet. When all turves have been removed the sheet is drawn clear away. Digging proceeds by placing the next spits on the second sheet now drawn to the 'face' edge of the pit (this also protects the top from trampling, etc.). When the pit has served its purpose the subsoil sheet may be drawn over the pit and emptied therein, to leave a perfectly clean sward behind. The fill-back is now tamped down and the turves are accurately replaced and beaten down. After good work no raw soil should be visible. In arable fields the cultivation layer is segregated and then replaced: in woodland the same care is taken with surface organic matter.

The appreciation by occupiers and particularly by graziers of this care on the part of the soil surveyor produces a respect most helpful to both parties.

The pit should be rectangular in plan, large enough for a man to sit comfortably in the bottom, and as deep as is necessary to expose the parent material. The position when possible should be oriented in such a way that the light shines into the end of the pit and indirectly illuminates both long sides at the same time. The tops of the long sides should be preserved so that no trampling or puddling occurs (or structure and constitution may be adversely affected). When digging pits into hill-sides it is particularly necessary for the profile record to prepare the *sides* running down hill. The 'cut-off' into the hill gives a false impression of the horizons by portraying them as approximately horizontal whereas in fact the true horizons will slope more or less as the slope of the hill.

The long sides of the pit may now be 'faced' and may be examined and recorded 'fresh', but as good structural and constitutional features do not usually appear until the soil has dried somewhat it is perhaps better to leave the whole pit for a while.[1] When the faces have dried out sufficiently one face may be picked with the fingers or a small knife blade so as to bring out the characteristics of the structural elements, while the other face may be cut back to show the soil matrix. During this latter operation the soil removed by each cut of the spade should be thrown out separately and carefully examined for structure, etc., as a corroboration of the detail to be observed in the picked face. It is noticeable that certain characteristics show up better in the loose fragments than in the picked face, for example, crumb and nutty structures are usually more clearly defined by the jolting of the mass of earth as it is tossed on to the ground, while prismatic or columnar structures are usually better seen *in situ*.

THE SOIL-PROFILE DESCRIPTION

The first operation in the examination of the soil profile is to divide it into its different layers or horizons, on any visible soil characteristics, and to give them names or numbers. Then proceed to the physical examination of the soil characteristics of each horizon by hand and eye. This is a matter of objective observation. Nothing, either visible or tangible, is without some significance in the make-up of the soil as a whole. Table 2 shows how the description of each layer may be laid out in an orderly manner.

TABLE 2

  I. Horizon or layer with depth and clarity of definition.
 II. Colour and its disposal (Munsell scale).
III. Texture (proportions of mineral constituents less than 2 mm).
 IV. Coarse skeleton (proportions and nature of stones).
  V. Structure (size and shape of soil aggregates).
 VI. Soil constitution (visible porosity and handling consistency).
VII. Organic matter (nature and distribution).
VIII. Roots (nature and distribution).
 IX. Drainage and water regime.
  X. Secondary or pedological deposits.
 XI. Distribution of carbonates and pH values.
XII. Fauna within the soil mass.
XIII. Scaled sketch of whole profile.

[1] Practical difficulties may arise in soils with high water-tables or on wet sites, but most of these may be overcome by means of sump holes and a pump. In the case of marshes and swamps, however, monoliths may be taken from the fresh pit and allowed to 'structure up' in the laboratory.

Item I. Horizon name: layers, horizons, and zones

In the horizonal differentiation of the soil profile the chief genetic divisions have the appearance of layers, and it is often convenient in the field to record the profile as Layer I, Layer II, etc., from the top to the bottom. When, however, it is possible to differentiate the layers by their genetic characteristics, use may be made of the horizon symbols: A, B, G, C, A–C, and subscripts 1, 2, 3, etc. Such symbols are usually applied to soils of temperate regions. Their use in the tropics has introduced more muddled thinking than it has prevented: they should be limited to temperate regions.

Horizons

The A horizon. A is the surface horizon which is in direct contact with climatic influences. Under the action of water it tends to lose soluble salts by drainage, and sometimes to lose fine particles of insoluble material by mechanical downwash. It is essentially a horizon of eluviation. In most soils the A horizon shows subdivisions within itself, and these are described by adding a numeral to the initial letter, starting with A_1 at the top of the mineral soil and proceeding downwards until the next genetic horizon is reached.

The A horizon is frequently overlaid by layers of vegetable debris (sometimes described as A_0) somewhat loosely named as humus layers, leaf litter, leaf mould, etc. Using the Swedish term *Förna* (see pp. 68–72) for such surface material, Hesselman has divided this into layers under the headings of F and H. In a manner similar to that of the division of the A horizon into A_1, A_2 etc., the F horizon may be subdivided into F_1, F_2, etc., according to the progressive decomposition of the annual leaf fall. When the botanical structure of the vegetation debris has been lost and the mass is truly humified, the horizon is designated as H. The manner in which the symbols F and H are used may be illustrated by the following description by Mattson and Ekman (1935) of the organic profile of a spruce stand:

F.00. Green needles picked from tree (i.e. living vegetation).
F.0. Dead needles shaken from tree.
F.1. 0–1 cm near trunk of tree.
F.2. 7–10 cm structure largely preserved.

F.3.   12–15 cm colour still light but structure partly destroyed.
H.   18–22 cm black, structureless humus.

Beneath the H layer the soil profile properly begins with the A_1 horizon which is dominantly of mineral origin.

The B horizon. B is essentially an illuvial horizon, or a horizon of accumulation of the materials derived from A. Such materials accumulate both by mechanical downwash and by chemical deposition. In certain soils, however, an accumulation horizon is formed from below by an upward current of salt-bearing water. Though illuvial, this is not usually termed a B horizon. Subdivisions of B are numbered as B_1, B_2, B_3.

The G horizon. The symbol is the initial letter of the word *Gley*.[1]

By the word 'gley' the people understand a more or less compact loamy or clay rock of a grey colour but less plastic and less sticky than usual. It frequently has a more or less clear, yet weak, greenish-blue tint.

The above quotation is from Vysotzky (1905). Since that time, however, the term has been used by pedologists to describe a specific set of soil-profile characteristics. The colours of the zone are produced by the alternate reduction and oxidation, hydration and dehydration of iron and manganese compounds by the fluctuating rise and fall of the ground-water zone. Pale blue, green, and yellow mottlings, as patches and streaks, occur *in* the soil *matrix* and/or *on* the *faces* of the structural elements. In many soils with sufficient clay to possess structure the elements of the Gley zone possess a structure peculiarly their own, which resembles a polygonal prism resting on its end. The inter-unit fissures may be inclined to the vertical. The faces possess an almost metallic lustre which, it is suggested, may be produced by polishing friction between the faces of the elements as they expand and contract on the rise and fall of the ground-water. It has also been suggested by Rode (1930) that the polish is caused by the downwash of fine clay into cracks formed during dry weather.

The characteristic Gley zone is not confined to any special depth in the profile. It is essentially a zone of alternating oxidation and reduction and, in consequence, is most frequently to be found in ill-drained, marshy, or meadow profiles.

The C horizon. C is the mineral material of geological origin in

[1] Pronounced to rhyme with 'play'.

which, under the influence of environmental factors, true soil horizons are forming as the soil develops.

The A–C horizon. In certain soils, particularly those derived from soft limestones, the rock weathers to form A and C horizons only, and there appears to be no development of a true B horizon. Calcium carbonate increases both in quantity and size of fragments from the top of A to the bottom of C. A horizon which has the properties of the mean of the A and C horizons may be found, and is termed the A–C horizon, since no horizonal boundaries can be determined. This is characteristic of Rendzinas and skeletal soils.

Kubiena (1953) has added several symbols to this profile notation to give more specific information about sub-horizons or zones. Such additions are valuable. A simple glossary of his terms is as follows.

A   to distinguish from
(A)   '*A bracket*', which describes raw soils that have not yet developed a true humic horizon but are 'through rooted by vegetation',
Ae   the *eluvial* or bleached layer,
A/Sa   the A layer enriched by salts,
B   to distinguish from
(B)   '*B bracket*', which describes a B layer *not* enriched by *illuviation* that has been formed by synthesis of clay *in situ*,
Bs   sesquioxide⎫
Bh   humus   ⎬ accumulation as in ortstein layers,
Ca   (preferably 'ca') represents a zone of calcium carbonate accumulation,
Mn   manganese dioxide zone,
Y   gypsum or calcium sulphate zone.

The 'Cr' nomenclature for tropical soils. Recent work in the tropics has shown that most tropical soils do not have horizons at all comparable with those for which the common A–B–C terminology was developed. The late C. F. Charter (1949) devised a scheme for the use of the soil surveyors of the Gold Coast which has been shown by H. Vine (1954) and P. H. Nye (1954) to be equally applicable to Nigeria.

In these countries, as over great parts of Africa, much of the material now composing the soil is not derived straight from the underlying solid rock but represents a mantle formed by a slow process of colluviation. Commonly the junction between the mantle or *creep layer* and the sedentary weathering or weathered rock is indicated by a *stone line* consisting of angular to sub-angular pieces of quartz, originating from quartz veins in the rock (Nye 1954). Termite activity and clay eluviation differentiate the mineral fractions so that the lower part of the creep layer is much more gravelly than the upper part. Superimposed on this are the effects of earthworms, which take the finest material from the subsoil and deposit it in casts on the surface. Since termites do not usually carry soil particles greater than 2 mm, and earth-worms do not appear to ingest anything greater than 0·5 mm, the resulting profile is characterized by the following attributes:

Cr W   Horizon formed from worm-casts,
Cr T   „   „   by termites,
Cr G   „   of gravel accumulation,
S_1 „   with little decomposing rock visible,
S_2 „   with many decomposing rock fragments,
S_3 „   of altered rock which still retains the structure of the fresh rock.

The Cr W layer usually shows a distinctly higher silt and clay content than the Cr T layer, while there is also an increase in these constituents in the Cr G layer. The layer containing the maximum clay content is usually that which lies immediately beneath the Cr G layer, i.e. the S_1 or S_2 horizons.

Layers. From what has previously been stated it is obvious that the genetic horizons A, B, and C, etc., cannot always be determined until after study in the laboratory. In field work therefore it is recommended that the different portions of the soil profile that may be recognized by direct observation are identified by numbers.

Depth and clarity of horizons and layers

Depth. The depth of each layer may be recorded in any convenient units but it is important that the unit is named. It is usually convenient to start measuring at the top of the pit and to work downward. Measurements over any one particular profile must be accurate, but it must be understood that within the same taxonomic unit the thickness of the various layers may vary somewhat. In the description of the modal profile, therefore, the degree of tolerance must be indicated. For example, if measurements of a horizon varied between 10 and 12 inches, the record would read: '11 in (10–12)'. The degree of tolerance permissible before a particular profile must be placed in another unit depends upon the instructions laid down for the survey. In the writer's experience, variations exceeding 15 per cent exceed the limits of tolerance.

Clarity. The clarity of the junctions between layers may be recorded as:

'Sharply defined', when the boundary occurs through not more than 1 in,
'Clearly defined', when the boundary occurs through not more than 2 in,
'Merging', when the junction changes gradually over a wide range.

The regularity with which the boundary runs horizontally may further be described as 'smooth', 'irregular', 'wavy', or 'tongueing'.

Item II. Colour and its disposal

Soils derive their colours from two sources:

(a) Organic matter; black, brown, grey.
(b) Mineral matter; iron: red, orange, yellow, brown, blue, green; manganese: black, brown.

Pinks, mauves, purples, and many other colours also occur, due to other minerals in the parent material, but these are not common or important. Various attempts have been made to correlate colour with fertility, but so far no very conclusive results have accrued.

The various coloured compounds of iron differ mainly in their degree of oxidation and hydration.

The colour range of soils over the main climatic zones tends to proceed from browns and yellows in the cool and humid zones to reds in the subtropical humid zones. The Red and Pink Insubrische soils of Switzerland on the southern slopes of the mountains are characteristic of such situations and, like the Terra Rossa, may be looked upon as the link between the Brown Soils of the temperate zone and the Red Soils of the subtropics.

The permanence of these colours depends to a certain extent upon the humidity of their environment. If certain red tropical soils are stored as monoliths in a cool humid environment they tend to lose some of their redness and to assume a yellowish tinge. Prolonged dehydration at a low temperature, however, tends once more to restore the sample towards its original colour.

In colour descriptions, it is helpful if the chief cause of the soil colour can be ascertained. For instance, a horizon may be black either from an accumulation of manganese dioxide or from humified organic matter, and a soil type may sometimes be defined on such evidence alone.

The angle at which light illuminates the soil may cause apparent colour changes due to reflection and absorption. The length of the shadows cast by the soil particles in the texture and structure of the soil mass also produces peculiar effects. Red soils tend to look redder in the afternoon than in the morning, while some soils may differ when looking uphill or downhill and towards the sun or away from it. Optical problems such as these form one of the chief difficulties yet to be overcome when mapping by eye from aircraft.

Because of the heterogeneity of most soil colours a perfect set of soil-colour standards has not yet been evolved, but a great advance in standardizing nomenclature has been made by the use of Munsell Soil Colour Charts. The Munsell notation is a specification of colour in terms of its three variables, called *hue*, *value* (brilliance), and *chroma*. *Hue* is the dominant spectral colour or its equivalent. It is the quality that distinguishes red from yellow-red, yellow-red from yellow, etc., and is related to the wavelength of the dominant component of the light. *Value* refers to apparent lightness as compared with absolute white, and is a function of the intensity of light. *Chroma* refers to the purity of hue, the saturation, or the apparent degree of departure from neutral greys or white. Neutral colours are those resulting from varying intensities of white light; they have no hue and zero chroma, and are the pure greys, black, and white.

The nomenclature for soil colour consists of two complementary systems, viz. colour names (which have been evolved from practice in the field) and the Munsell notations. For the proper recording of colour both must be used, and the moisture status of the soil must be described at the same time. A description could read:

0–6 in Plough layer, greyish-brown (10 YR 5/2 dry: 10 YR 4/3 moist) - - - - -.

Excellent as this system is, the soil man still has to learn his own technique in the field, for example, the authors always record their wet colour at 'sticky point', while determining field 'texture'.

The usual method of determining the colour is to hold a small fragment of the soil behind the colour card and to move it about behind the apertures on the card until the best colour match can be recorded. In addition to recording what may be called the 'basic colour', notes should be taken of the way in which the basic and other colours are distributed. Colours tend to occur in some sort of order and may be recorded in simple terms such as the following:

(a) Self-coloured. The whole horizon is homogeneous.
(b) Speckled. Small particles of various colours fairly uniformly distributed in an otherwise self-coloured matrix.
(c) Streaky. Colourings are dominantly vertical.
(d) Rippled. Very small, dominantly horizontal markings.
(e) Waved. Large, dominantly horizontal markings of an undulating nature.
(f) Banded. Large and generally flat and even.
(g) Cloudy. Large ill-defined patches which merge without definite boundaries.
(h) Mottled. Multicoloured particles in a generally cloudy mass. Entirely different from 'speckled'.

The masking of one colour by another is common in organic soils, and it may be convenient to oxidize the organic matter with hydrogen peroxide and record the colour of the residual mineral material to verify a suspected translocation of iron, though this is obviously not a usual field technique.

Item III. Texture

The term texture has been used in the past in a loose manner to include a wide range of soil characteristics. It now refers specifically to the inorganic solid material into which the soil mass may be dispersed without comminution. Texture defines the relative amounts of coarse and fine material present in the soil mass and must not be confused with its structure or constitution. The terms 'clay', 'sand', etc., in a texture description refer to the dominance of particles within a certain size-range, and bear no relation to any other characteristic. It is important to attempt to assess the texture of the mineral portion of the soil without regard either to the degree of aggregation or to the quantity of organic matter present.

Table 3 indicates some systems of texture grades.

The British system of texture classes

The soil survey of Great Britain has now revised its classification of soil texture classes to conform fairly closely to that in use in the United States (United States Department of Agriculture 1951).

The properties of the grades that determine the texture of a soil are as follows:

Coarse sand consists of grains between 2 and 0·2 mm in diameter that are large enough to grate against each other, and can be detected individually both by feel and sight.

Fine sand consists of grains between 0·2 and 0·02 mm and the grading is therefore much less obvious, but can still be detected, although individual grains are not easily distinguished by either feel or sight.

Silt. Individual grains of silt cannot be detected, but silt feels characteristically smooth and soapy, and only very slightly sticky.

Clay is characteristically sticky, although some dry clays require a great deal of moistening and working between the fingers before they develop their maximum stickiness.

TABLE 3

Some systems of texture grades

(in mm)

The international scale

Clay	Silt	Sand		Gravel
		Fine	Coarse	

0·002   0·02   0·2   2

U.S. Department of Agriculture

Clay	Silt	Sand	Gravel

0·002   0·05   2

U.S. Bureau of Soils and Public Roads Administration

Clay	Silt	Sand		Gravel
		Fine	Coarse	

0·005   0·05   0·25   2

Massachusetts Institute of Technology and British Standards Institution

Clay	Silt			Sand			Gravel

0·002   0·006   0·02   0·06   0·2   0·6   2

Soil texture class descriptions. The following list of soil texture classes is taken from the *Field handbook of the Soil Survey of Great Britain* (1960).

Sand. Soil consisting mostly of coarse and fine sand, and containing so little clay that it is loose when dry and not sticky at all when wet. When rubbed it leaves no film on the fingers.

Loamy sand. Consisting mostly of sand but with sufficient clay to give slight plasticity and cohesion when very moist. Leaves a slight film of fine materials on the fingers when rubbed.

Sandy loam. Soil in which the sand fraction is still quite obvious, which moulds readily when sufficiently moist but in most cases does not stick appreciably to the fingers. Threads do not form easily.

Loam. Soil in which the fractions are so blended that it moulds readily when sufficiently moist, and sticks to the fingers to some extent. It can with difficulty be moulded into threads but will not bend into a small ring.

Silt loam. Soil that is moderately plastic without being very sticky, and in which the smooth soapy feel of the silt is the main feature.

Sandy clay loam. Soils containing sufficient clay to be distinctly sticky when moist, but in which the sand fraction is still an obvious feature.

Clay loam. The soil is distinctly sticky when sufficiently moist, and the presence of sand fractions can only be detected with care.

Silty clay loam. This contains quite subordinate amounts of sand, but sufficient silt to confer something of a smooth soapy feel. It is less sticky than silty clay or clay loam.

Silt. Soil in which the smooth, soapy feel of silt is dominant.

Sandy clay. The soil is plastic and sticky when moistened sufficiently, but the sand fraction is still an obvious feature. Clay and sand are dominant, and the intermediate grades of silt and very fine sand are less apparent.

Medium clay. The soil is plastic and sticky when moistened sufficiently and gives a polished surface on rubbing. When moist the soil can be rolled into threads. With care a small proportion of sand can be detected.

Heavy clay. Extremely sticky and plastic soil, capable of being moulded when moist into any shape and taking clear fingerprints.

Silty clay. Soil that is composed almost entirely of very fine material but in which the smooth soapy feel of the silt fraction modifies to some extent the stickiness of the clay.

Qualifying adjectives may be added to certain of the above descriptions, for example, loamy *coarse* sand, *fine* sandy loam, etc.

Soils containing appreciable quantities of organic matter may be described as:

Slightly humose. Soils with the feel of the mineral texture grades but containing about 8–13 per cent organic matter.

Humose. Soils with the feel of the mineral texture grades but containing about 13–25 per cent organic matter.

Very humose. Soils with a loamy feel which are moderately plastic when moist and will roll into short threads. Containing about 25–40 per cent organic matter.

Organic soil. Dark-coloured soil without plasticity and incapable of forming threads when rolled between the fingers. Smooth, light and often powdery to the feel. More than about 40 per cent organic matter.

High contents of organic matter tend to make sandy soils and clay soils feel more loamy. Organic matter feels rather like silt, but whereas silt leaves a smooth soapy surface when smeared between the finger and thumb, organic matter tends to give a frayed appearance. Finely divided calcium carbonate also gives a silt-like feeling to the soil.

To assess texture, the soil is moistened until it reaches its maximum stickiness or plasticity, and a small amount is rubbed down between fingers and thumb. (It is very important to do this thoroughly in order to eliminate the masking effects of structure and consistence.) An attempt is then made to estimate the relative proportions of coarse sand, fine sand, silt, and clay, and the soil is given its texture name according to the proportions of these groups present.

For field work in general it is always better to work for a time with an experienced field-man to pick up the technique, but when this is impossible the writers have found the following method reasonably successful in the early stages. The student first becomes familiar in the laboratory with the 'feel' of certain well-defined and standard grades. Mechanical analysis figures are studied and a number of soils are selected in which fine sand and silt add up to about 50 per cent. The 'feel' of these soils is determined, and the modifying influence of more or less clay or more or less coarse material is examined and graded into as many divisions as possible. Modification in 'feel' produced by organic matter, chemical deposits, skeletal material, or stones should also be studied, since these reactions are much more marked in the field than in the laboratory. After some experience a high degree of accuracy may be attained.

Another interesting method for the field classification of texture has been evolved by Krasiuk (1929). An outline of the scheme is given in Table 4. The effect of texture varies greatly with the type of soil under observation. Indirectly or directly, however, texture affects structure and constitution and, in consequence, influences aeration, permeability, drainage, and finally root development and penetration. At the same time the physical stability, liability to erosion, creep, or slip may also be affected.

Item IV. The mineral skeleton: nature of stones, etc.

This item, though of very great importance in the field, is rarely considered at all in the laboratory examination of a soil sample (other than as a soil monolith), because in preparing the sample for analysis the stones are usually removed. It is important, therefore, that fairly full notes should be taken in the field. A record must be made of the absence or presence and, if present, the quantity and distribution through the soil mass, of the unweathered or partially weathered mineral material greater than 2 mm diameter. Soil behaviour may be considerably affected by varying quantities of such material, which may conveniently be described as 'stones' and classified under four main headings according to chemical nature, shape, size, and quantity.

Chemical nature

(a) Residual parent material or adventitious material that is *capable* of further weathering and comminution to produce 'fine earth'. Such material may have both chemical and physical effects upon the soil-forming processes and is characteristic of immature soils.

(b) Residual or adventitious materials *incapable* of further chemical decomposition as long as present conditions prevail. They may, however, be subjected to further comminution, but their only direct effect upon the soil mass is a physical one. Such materials as quartz pebbles and flints may be included in this category.

(c) Concretionary materials which, through chemical or physico-chemical reactions, have accumulated in the soil mass, though not usually referred to as 'stones', frequently behave as such. Their description and properties are more properly dealt with in the section on secondary chemicals.

TABLE 4
Textural types (from Krasiuk 1929)

Mechanical subdivision	Feel between fingers	Appearance under lens	Dry state	Wet state	Rolling between fingers	Finer subdivisions
I. Clays	Fine homogeneous powder	Large grains of sand absent	Very compact—forms crumbs which are very hard	Very sticky plastic	Gives long threads	Acc. to plasticity: (a) heavy, (b) light
II. Loams	Not quite homogeneous powder	Among clay, sand particles visible	Compact—crumbly but not so hard	Plastic	Gives threads with difficulty and these will not bend into ring	Acc. to amount of sand: (a) heavy, (b) medium, (c) light
III. Sandy loams	Heterogeneous, loam alternates with sand (one hor. sand other loam)	Clayey part mixed with sands	Not uniform in compactness	Slight plasticity	Threads form with great difficulty	Acc. to size of sand grain: (a) coarse sandy, (b) fine sandy
IV. Loamy sands	Sandy particles predominate — small admixture of clay	Sandy particles predominate — small	Dries into ill-defined crumbs from surface of which sand is easily rubbed	Cannot be rolled into threads		Acc. to size of sand grains: (a) large grained, (b) fine, (c) loess-like
V. Sands	Consist almost exclusively of sand grains			Form a flowing liquid mass		(a) clayey sands (b) friable sands
VI. Gravelly or stony	Together with clayey and sandy particles contains large number of lumps of rock in form of gravel (3–10 mm) and stones (> 10 mm)					Acc. to amount and composition of fine earth may be clayey, loamy, loamy sand, or sand

The recognition of the constituents of either of the above categories is merely a matter of the examination of a hand specimen and possibly of its fracture by a blow with a hammer.

Shape

Shape may be divided into

(a) Angular (including cubic and flat varieties),
(b) Sub-angular (including cubic and flat varieties),
(c) Rounded (including nodular formations),
(d) Shaly (cleaved structure),
(e) Tabular.

Size

The length in inches of the *major* axis is always understood to be taken unless otherwise stated.

The size is always given by name (e.g. coarse gravel).

Gravel	Coarse gravel	V. small stones	Small stones	Medium stones	Large stones	Boulders
$\frac{1}{8}''-\frac{1}{4}''$	$\frac{1}{4}''-\frac{1}{2}''$	$\frac{1}{2}''-1''$	$1''-2''$	$2''-4''$	$4''-8''$	$>8''$

It is thus possible to describe a stony soil as containing 'sub-angular, coarse gravel', or 'rounded, large to medium stones', etc.

Quantity

The quantity of stones present may be recorded under one of the following headings.

(a) Stoneless or nearly so.
(b) Slightly stony, not sufficient to interfere with cultivation.
(c) Very stony, enough to interfere with cultivation and visibly to act as a serious diluent to the soil mass 'fine earth'.
(d) Occasional boulders.
(e) Live rock exposed.
(f) Rock or stones dominant.

The general effects produced by stones in the soil mass depend to a great extent upon their nature, quantity, and distribution. For example, a generous proportion of stones on the surface of the soil may act as a mulch and tend to prevent losses by evaporation; excessive stoniness, on the other hand, will tend to produce a too loose constitution and cause insufficient water retention. In arable practice, too many stones will reduce the effective volume of the soil available for rooting. Local or micro-leaching may occur in soil with excessive quantities of stones of acid rocks to produce a degradation that would not otherwise occur; frequent examples of this action are to be found among flint and siliceous gravel soils. If, however, the stones are derived from calcareous rocks, the rate of soil degradation is greatly reduced owing to the maintenance of the base status by solution of the basic material, as in the Rendzinas and Red and Brown calcareous soils of the temperate zone.

The distribution or the position of the stones in the soil mass may have a marked effect upon the utilization of the site in that a layer of gravel in the 'subsoil' of a clay soil may be sufficient to allow efficient drainage and economic utilization that could not otherwise occur. A clay 'subsoil' under a gravelly surface layer, on the other hand, may produce an imperfectly drained profile of little value.

Item V. Structure

Soil structure is an exceedingly interesting phenomenon, and is capable of investigation under two different heads and in two distinct ways.

Soil structure can be studied in the field by the simple exercise of the senses of touch and sight, and descriptions so obtained refer to the soil *macro*-structure. What cannot be determined in the field because of the extremely small size of the aggregates and their intricacy is termed *micro*-structure or fabric. It is bound up with soil life and fertility. The ultimate *effects* of micro-structure are frequently visible in the field by their effects upon the macro-structure and soil profile generally, but the study of micro-structure itself is still a problem for the pedologist in the laboratory.

Structure therefore, is the term defining the aggregates or fragments as *seen* in the field, into which the soil mass will crack and break on the natural drying of an exposed face, or when a spadeful of the soil is tossed about a metre into the air and allowed to fall into fragments by shock. These fragments are usually termed *structural elements* or peds and are aggregates of the textural elements. No kind of structure is fortuitous; each is quite characteristic of a particular soil fabric and is formed as the result of the operation of certain definite factors. Every major soil group possesses a specific structure which, under the influence of a new set of factors, will tend to change into some other equally characteristic type.

Structure influences the soil in almost all of its reactions, but especially with regard to aeration, moisture, heat, permeability, and water capacity. To some extent it is related to liability to erosion.

Biotic influence. The development of a *natural* soil structure is due to the *inherent* biological environment of the soil, and unless interfered with a state of equilibrium will persist. The lesser animals, particularly the burrowing species by their manipulation of soil particles, and the worms, insects, and their grubs with their peculiar cementing secretions and their excreta, all tend to foster the development of crumb structure. By penetration and root pressure, root development is frequently responsible for the size and shape of structural elements. An excellent example of such development of crumb structure may be seen in the upper (sod) horizons of old grasslands. Also some degree of lamination may be observed under forest where wind vibration causes stresses upon the lateral root systems of large shallow-rooted trees. Podzols also frequently exhibit lamination in the A_2 horizon, but this is probably due to the method of decomposition of the humus and to its reaction on the mineral mass.

Artificial structure may be induced by implements, the application of chemicals, or water control. For example, poorly structured podzols may be made into crumb structured arable soils by the application of suitable fertilizers and cultivations. Intractable clay soils, by skilful cultivation and manuring, may be made to acquire the physical properties and crumb structure of a loam. The aim, in fact, of all rational cultivation is the attainment of such an 'agronomic' crumb structure.

Unnatural pressure due to man's careless usage of heavy implements or too violent operations during cultivation leads to the destruction of the tilth or agronomic structure by packing and reduction of pore space. Then follows the typical agricultural pan with its impeded aeration and drainage.

Unnatural pressure invariably exerts a deleterious effect upon the size and shape of structural elements as well as upon their consistence, so that the observer should find little difficulty in diagnosing it in the field.

Soils low in colloid content, which even under the influence of flocculating factors do not bind into definite aggregates, are frequently referred to as being of 'single grain structure'.

Structural features. Structural units belong to one or other of two main classes. The first occurs in the surface soil and is primarily developed under the influence of vegetation, fauna, and the direct effects of climate. The second class occurs in the subsoil; while it is usually insulated from the direct effects of the atmospheric climate and the surface fauna, it is frequently affected by vegetation, and is always influenced by the percolation stream. The progressive increase in size of the structural elements with depth that is found to be characteristic of Brown Forest Soils is probably due, among other things, to the diminution of humus down the profile and to the presence of a more highly basic ground-water.

Causes of special shapes. The factors thus described that govern structure development are fairly well known, but little or nothing is known about the causes of the specific shapes produced in different soil types. From a study of the measurements of the shrinkage of naturally moist structural elements the author believes that there is some definite orientation of the ultimate soil particles in a soil mass that produces a form of cleavage plane along which the elements tend to separate. If fresh well-shaped elements of the prism or column-like type are allowed to dry out naturally, it is found that there is a greater proportional shrinkage parallel to the major axis than along the parallels of the minor axes. Also, if and when columnar and column-like structures disintegrate, they invariably fracture along the parallels of the horizontal axes. Such spaces as exist between the elements will be closed up more quickly through the elongation of the vertical faces by the percolating water than will the rest of the faces of the element so that a strain will be exerted whereby the element is attempting to expand laterally but is locked by friction. The result is a shearing force which cracks the elements along the parallels of the minor or horizontal axes.

If a certain humus content and base status can be responsible for a certain orientation of ultimate particles it should be possible, by changing this base status and colloidal binding power by leaching or regrading the soil mass, to change the shape and size of the structural elements. This is in fact the case and can be observed both in the laboratory and in the field. Field observations of the reclamation of a salt marsh to a neutral grassland (i.e. a sodium clay to a calcium clay) show that the process is not rapid, but proceeds firstly by a degradation of column-like structure to jointed columnar, then to a somewhat spheroidal cube (rounded edges and corners), and finally to the true cube to produce a crumb surface soil with a nutty subsoil. The stability of the cube type is probably due to the fact that the shrinkage takes place evenly along all parallels.

Description of aggregates

There are three general methods for the description of soil aggregates. The earliest is that evolved by the Russian pedologist Zakharov which describes *structure* strictly in terms of the *size* and *shape* of the aggregates. The *consistence* or the *tenacity* of the aggregates is described separately.

The Russian system is shown in Table 5. There are three *types* of structural elements classified according to their *main shape*. The *types* are then further divided into nine *kinds* according to the clarity of the definition of the *faces* and *edges* of the elements, and then these are further subdivided into numerous *varieties* according to the *size* of the elements.

The second system has been evolved by the U.S. Soil Survey (United States Department of Agriculture 1951). This system classifies *peds* into *types* by shape and *classes* by size, much as in most other classifications, but a third observation is also recorded: the *grade* of structure. *Grade* describes the clarity of definition of the peds or the degree of structure development, either in the soil mass or after displacement, correlated with their resistance to disruption or displacement.

Degree of structural development. This is distinguished in the field by the proportion of the soil appearing as aggregates and by the development of aggregate faces. Peds that are stable and persistent show faces that are well formed and are usually smoother than the interior when the aggregate is broken across, and they sometimes have a different hue. The faces may be stained with a coating of other material or have an abundance of roots running along them. Very well formed aggregates may possess smooth shiny faces.

Terms for the grade or degree of structural development are as follows.

Structureless. No observable aggregation or no definite orderly arrangement of natural lines of weakness. *Massive if* coherent; *single grain* if non-coherent.

Weakly developed. Poorly formed indistinct peds that are barely observable in place. When disturbed, the soil breaks into a mixture of a few entire peds, many broken peds, and much unaggregated material. If necessary for comparison, this grade may be subdivided into *very weak* and *moderately weak.*

5

TABLE 5

Russian system of soil structure classification

(From Zakharov 1927; revised translation by Muir 1934)

Type	Kind	Varieties	Dimensions
I. Cube-like structure — soil fragments equally developed along the three axes	A. Faces and edges feebly manifested, soil aggregates mostly complex and irregular in shape		
	1. Clod structure	Large cloddy	> 10 cm
		Small cloddy	10–5 cm
	2. Crumb structure	Large crumbs	5–3 cm
		Med. crumbs	3–1 cm
		Small crumbs	1–0·5 cm
		Pulverescent	< 0·5 cm
	B. Faces and edges more or less clearly manifested, aggregates well defined		
	3. Nutty structure	Cuboid	> 20 mm
		Large nutty	20–10 mm
		Med. nutty	10–7 mm
		Small nutty	7–5 mm
	4. Granular structure	Large granular	5–3 mm
		Med. granular	3–1 mm
		Small granular (powder-like)	1–0·5 mm

			Length of vertical axis

II. Prism-like structure — soil fragments predominantly developed along the vertical axis

- **A.** Faces and edges indistinctly manifested, aggregates complex and unclearly defined
  - **5.** Column-like structure

	Length of vertical axis
Large columns	> 5 cm
Med. columns	5–3 cm
Small columns	< 3 cm

- **B.** Faces and edges distinctly manifested, aggregates more or less well defined
  - **6.** Prismatic structure, with uniform, even often shining, surfaces and sharp edges

	Length of vertical axis
Large prisms	> 5 cm
Prismatic	5–3 cm
Small prisms	< 3 cm
Prismatic pencil-shaped (more than 5 cm in length)	1 cm

  - **7.** Columnar structure — the upper end ('top') rounded and the lower one flat

	Length of vertical axis
Large columns	> 5 cm
Columnar	5–3 cm
Small columns	< 3 cm

III. Plate-like structure — soil aggregates predominantly developed along the two horizontal axes

	Thickness

- **8.** Platy structure — horizontal 'planes of cleavage' more or less developed

	Thickness
Schistose	> 5 cm
Platy	5–3 cm
Laminar	3–1 cm
Foliated	< 1 cm

- **9.** Squamose structure—horizontal faces comparatively small, partly curved

	Thickness
Shelly	> 3 cm
Coarsely squamose	3–1 cm
Finely squamose	< 1 cm

Note. Large column-like fragments are sometimes called pedestals

Moderately developed. Well formed distinct peds that are moderately durable and evident but not distinct in undisturbed soil. The soil material, when disturbed, breaks down into a mixture of many distinct entire peds, some broken peds, and little unaggregated material.

Strongly developed. Durable peds that are quite evident in undisplaced soil, adhere weakly to one another, and become separated when the soil is disturbed. The soil material consists very largely of entire peds and includes a few broken peds and little or no unaggregated materia. IfInecessary for comparison, this grade may be subdivided into *moderately strong* and *very strong*.

Table 6 shows the American classification for structure from the *Soil survey manual* (United States Department of Agriculture 1951).

Grades of structure do not adequately cover all the characteristics of structure since no allowance is made in the description for the proportion of air spaces to solids in the soil mass. Certain Australian surveyors use the term *grade* as a quantitative measure of the proportion of stable 'peds' (structural aggregates) left after a spadeful of the soil has been tossed down and shattered; it is recorded as a percentage from visual estimation only.

A third system of description is given in Table 7.

Sizes are graded primarily into English units and their nearest metric equivalents are given as a secondary item. The nomenclature differs from the Russian system in many important details, so that the names of particular shapes are much more in keeping with their general appearance; for example, the name *nutty* applies to something which resembles a nut (i.e. it is rounded). Also, an effort is made to describe some shapes not catered for in either of the other systems.

Soil porosity

This depends upon the size, shape, and quantity of visible spaces throughout the whole soil mass. They may contain air or water or both and they occur as cracks and cavities both within the structural aggregates and between them. Porosity is a natural field characteristic, influenced by flora and fauna as well as by the physical properties of the soil material itself, and so it bears little or no relation to laboratory determinations of soil 'pore space' as made on a 'prepared fine earth' sample.

Causes of soil porosity. The *chief* cause for the cracking of soils is the shrinkage of colloids on dehydration.

TABLE 6

U.S. system of soil structure classification

TYPE (shape and arrangement of peds)

Class	Plate-like with one dimension (the vertical) limited and greatly less than the other two; arranged around a horizontal plane; faces mostly horizontal	Prism-like with two dimensions (the horizontal) limited and considerably less than the vertical; arranged around a vertical line; vertical faces well defined; vertices angular		Block-like; polyhedron-like or spheroidal, with three dimensions of the same order of magnitude, arranged around a point				
				Block-like; blocks or polyhedrons having plane or curved surfaces that are casts of the moulds formed by the faces of the surrounding peds		Spheroids or polyhedrons having plane or curved surfaces which have slight or no accommodation to the faces of surrounding peds	
				Faces flattened; most vertices sharp angular	Mixed rounded and flattened faces with many rounded vertices	Relatively non-porous peds	Porous peds
	Platy	Without rounded caps	With rounded caps				
		Prismatic	Columnar	(Angular) blocky[1]	Sub-angular blocky[2]	Granular	Crumb
Very fine or very thin	Very thin platy; < 1 mm	Very fine prismatic; < 10 mm	Very fine columnar; < 10 mm	Very fine angular blocky; < 5 mm	Very fine sub-angular blocky; < 5 mm	Very fine granular; < 1 mm	Very fine crumb; < 1 mm
Fine or thin	Thin platy; 1–2 mm	Fine prismatic; 10–20 mm	Fine columnar; 10–20 mm	Fine angular blocky; 5–10 mm	Fine sub-angular blocky; 5–10 mm	Fine granular; 1–2 mm	Fine crumb; 1–2 mm
Medium	Medium platy; 2–5 mm	Medium prismatic; 20–50 mm	Medium columnar 20–50 mm	Medium angular blocky; 10–20 mm	Medium sub-angular blocky; 10–20 mm	Medium granular; 2–5 mm	Medium crumb; 2–5 mm
Coarse or thick	Thick platy; 5–10 mm	Coarse prismatic; 50–100 mm	Coarse columnar; 50–100 mm	Coarse angular blocky; 20–50 mm	Coarse sub-angular blocky; 20–50 mm	Coarse granular; 5–10 mm	
Very coarse or very thick	Very thick platy; > 10 mm	Very coarse prismatic; > 100 mm	Very coarse columnar; > 100 mm	Very coarse angular blocky; > 50 mm	Very coarse sub-angular blocky; > 50 mm	Very coarse granular; > 10 mm	

[1] Sometimes called *nut*. The word 'angular' in the name can ordinarily be omitted.
[2] Sometimes called *nuciform*, *nut*, or *sub-angular nut*. Since the size connotation of these terms is a source of great confusion to many, they are not recommended

TABLE 7

Clarke's system of soil structure classification

General appearance	Appearance definition	Name	Sizes (English)	Sizes (Metric)
Cubic	Well-defined cubes	Large cubic	>6 in	>15 cm
		Med. cubic	6–2 in	15–5 cm
		Small cubic	2–1 in	5–2·5 cm
	Ill-defined cubes	Large (A or R) cloddy	>6 in	>15 cm
	Angular (A)	Med. (A or R) cloddy	6–2 in	15–5 cm
	Rounded (R)	Small (A or R) cloddy	2–1 in	5–2·5 cm
	Conchoidal fragments	Large starchy	>¾ in	>2 cm
		Small starchy	<¾ in	<2 cm
	Roughly rounded *solids* with good air spaces	Large nutty (walnut)	1 in	25 mm
		Med. nutty (filbert)	1–½ in	25–12 mm
		Small nutty (peanut)	½–¼ in	12–6 mm
	With few air spaces	Gunshot	¼–⅛ in	3–1 mm
		Large granular	⅛–1/16 in	<1 mm
		Small granular	<1/16 in	
	Roughly rounded *aggregated* small particles with well-defined air spaces in aggregates	Large crumb	3/8–¼ in	9–6 mm
		Med. crumb	¼–⅛ in	6–3 mm
		Small crumb	⅛–1/16 in	3–1 mm
		Crumb dust	<1/16 in	<1 mm

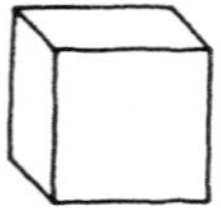

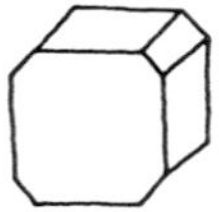

TABLE 7 (*contd.*)

Clarke's system of soil structure classification (*contd.*)

General appearance	Appearance definition	Name	Sizes: English	Sizes: Metric
Prismatic 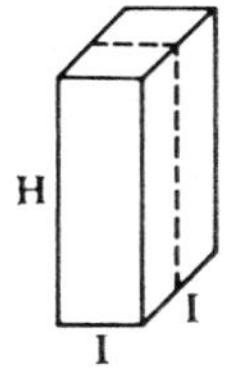	Well-defined prisms	Large prismatic	$> 2\times2\times$H (in)	$> 5\times5\times$H (cm)
		Med. prismatic	$2\times2\times$H to $1\times1\times$H	$5\times5\times$H to $2{\cdot}5\times2{\cdot}5\times$H
		Small prismatic	$< 1\times1\times$H	$< 2{\cdot}5\times2{\cdot}5\times$H
Columnar 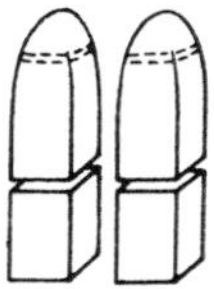	Well-defined prisms with *indefinite tops*, i.e. columnar	Large columnar Med. columnar Small columnar	Sizes as prismatic	
	Jointed columns Series of prisms, usually massive with *wide* vertical cracks and narrow horizontal cracks	Large *jointed* columnar Med. *jointed* columnar Small *jointed* columnar	Sizes as prismatic State height of column and cross-section at top and base	
Laminated 	Plates (flat)	Slabby Platy Foliated	$> \frac{1}{2}$ in H $> \frac{1}{8}$ in $< \frac{1}{8}$ in	> 13 mm > 3 mm < 3 mm
	Scales Curved (saucer-like)	Scaly Flaky	$> \frac{1}{8}$ in $< \frac{1}{8}$ in	> 3 mm < 3 mm

TABLE 7 (contd.)

Clarke's system of soil structure classification (contd.)

In addition to the systematic classification of the main types of structural elements, certain soils may be found wherein aggregates occur which do not readily allow of such simple description, and a further group is needed.

Pyramidal (tetrahedra)	Large pyramids Med. pyramids Small pyramids Very small pyramids	> 6 in sides 6–2 in 2–1 in < 1 in	> 15 cm 15–5 cm 5–2·5 cm < 2·5 cm
Inverted pyramidal	Large pyramids Med. pyramids Small pyramids Very small pyramids	Sizes as above	
Polyhedral State number of faces	Large polyhedrons Med. polyhedrons Small polyhedrons	> 1 in 1–½ in < ½ in	> 2·5 cm 2·5–1·2 cm < 1·2 cm
Single-grained or structureless	Defined generally by 'feel' Single-grained Mealy (sharp sand with hydrated sesqui- oxides) Powdery Desert dust Volcanic dust		2–1 mm 1–0·2 mm < 0·2 mm

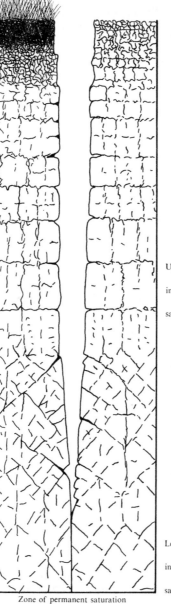

Very humic layer,

crumb structure,

to

less humic cubic,

merges to

jointed prismatic

with little or no

humus, merging

to

large prismatic

with

no humus

Upper fringe of

intermittent water

saturation

Wetter zone

Prismatic elements

crack along and at

right angles to

face of shrinkage,

changing to

pyramidal elements

with polished faces

Lower fringes of

intermittent water

Gley

saturation

Zone of permanent saturation

FIG. 6

Imagine an ordinary Brown Forest Soil under grass during a hot summer in England. (The following is based on many observations made by the author on clay profiles during summer droughts.) The sod layer rapidly dries out under direct insolation aided by transpiration. Root channels form the first lines of weakness in the fabric; a surface crack develops. Soils tend to crack perpendicularly to their drying surfaces so that the first crack is a vertical one. This incipient crack, once started, exposes at least two more faces for desiccation, and allows horizontal cracks to develop in sequence as the initial crack deepens. Since there is a higher humus content nearer to the top of the soil, shrinkage there is greater than it is lower down. The first result is the development of great numbers of vertical and horizontal cracks separating structural aggregates mainly of the cubic type. This allows grass roots to run in all directions to give rise to weakness lines, and thus open new faces obliquely. The sod layer is also characterized by countless numbers and varieties of soil fauna, which tend to aerate the soil further with their channels, or weakly to agglomerate the soil particles into crumbs or grains with their excrements. All together these operations are responsible for the perfect crumb structure and highly porous (mellow) constitution of the sod layer. With increasing depth the shrinkage becomes less, cracks appear less frequently and become dominantly vertical and horizontal because the grass roots are sparser or even absent, while the activities of small fauna are also reduced. The A horizon therefore is characterized by small cubic or prismatic aggregates with well-defined vertical and horizontal cracks between them. On further drying, the fissures between these structured aggregates allow more evaporation from the B horizon. This causes the dominant vertical fissures therein to continue to deepen, but because of a lower humus content and a lower population of fauna, crumb formation is inhibited. Cracking is almost entirely vertical, while the subsequent horizontal cracks become progressively fewer as evaporation from the free vertical face decreases (Fig. 6).

When rain falls some of the finer material, both organic and inorganic, from the soil surface is washed down the vertical cracks. As the structural aggregates expand this material becomes firmly trapped and may give the faces and edges of the aggregates a coating of material much more reactive to wetting and drying than the inner portion of the aggregates. In this way the initial pores of the soil fabric become remarkably stable, and fissures between both vertical and horizontal faces may be found in the same places year after year.

Certain soils exhibit another kind of cracking phenomenon under the influence of a high water-table. This occurs in the zone of imperfect drainage. It is associated with incipient gleying in the zone of alternate reduction and oxidation where ground-water falls or rises in the soil fabric. This zone never completely dries throughout but it does dry selectively from the fissures around the large prismatic aggregates. The lower and wetter of these aggregates, because of the soil pressure above them and the comparatively incompressible material beneath, tend to fracture obliquely at an angle of about 45° to the major vertical fissures. Subsequently further dehydration causes cracks to develop perpendicular to these new faces so that this zone of selective and partial dehydration gives rise to the characteristic prismatic structure so typical of clay soils under temperate maritime climates.

The forms of cracking so far described are not the only causes of air spaces in soil. Holes may result from the selective eluviation of fine or soft materials out of a heterogeneous matrix, or by the selective leaching of chemical substances either by complete drainage or to some other part of the profile. Finally, but by no means least, are the activities of all sorts of fauna and flora.

The description of porosity in the field proceeds under two headings,

(a) the nature of the spaces within the soil aggregates,
(b) the nature of the spaces surrounding the soil aggregates.

Aggregate porosity. This is defined as follows:

Fine porous < 1 mm diam.
Porous   1–3 mm diam.
Spongy   3–5 mm diam.
Cavernous 5–10 mm diam.

Nature of spaces between aggregates. The shape of gaps may bear some relation to the shape of the solid aggregates, for example, true rectangular prisms will usually be separated by vertical-sided fissures, but not necessarily. Any terms adequate to describe their distribution and shape are allowable. If holes are due to floral or faunal influence, they should be described as such. Sizes are recorded as:

Fine fissured   < 1 mm wide
Fissured   1–3 mm wide
Wide fissured   3–5 mm wide
V. wide fissured < 5–10 mm wide

Item VI. Constitution

Soil compactness

This is related to soil constitution and describes the resistance of the soil fabric to penetration or disintegration. In the field it is usually assessed and described by the simple reactions to digging, augering, or picking with a pocket-knife.

Description	Spade	Auger
Very compact	will not enter; pick or bar needed	virtually unborable
Compact	enters with difficulty; much fragmentation	good bite, fills threads with broken aggregates
Loose	enters easily and spit falls readily into pieces	particles fall out of threads and run through fingers
Friable	digs well with fine fragmentation	comes up loosely packed
Indurated (implies cementation)	pick needed to break, then digs easily	spins and grinds soil to dust
Tenacious	clogs and tears away from uncut faces	sucks noisily but comes up full

Handling consistency

This refers to the ability of a fragment of the soil to resist disruption or deformation in the hand. The American (United States Department of Agriculture 1951) system of classification describes the soil at three specific moisture contents. It is a practical method that is capable of giving a great deal of useful information.

The first investigation is made when the soil is at or just above its *field capacity*. (Field capacity may be defined as 'the amount of water held in a soil after the excess gravitational water has drained away and after the rate of downward movement of water has materially decreased'.) The observation is recorded under the heading of *Consistence when wet*, and two attributes are defined:

(a) *Stickiness*, the quality of adhesion to other objects. There are four degrees of stickiness.

Non sticky
Slightly sticky
Sticky
Very sticky
} The field diagnosis is fully described in the *Soil survey manual* (United States Department of Agriculture 1951)

(b) *Plasticity*, the ability to change shape continuously under the influence of an applied stress and to retain the impressed shape on removal of the stress. There are four degrees of plasticity.

Non-plastic
Slightly plastic
Plastic
Very plastic
} The field diagnosis is fully described in the *Soil survey manual*

Consistence when moist. This attribute is determined when the moisture content is about half-way between 'field capacity' and 'air dry'. It is assessed in the field by attempting to crush in the hand a mass of soil which appears to be 'slightly moist': the soil should *feel* moist and yet change its colour on further wetting. There are five degrees for this form of consistence.

Loose
Friable
Firm
Very firm
Extremely firm
} The field diagnosis is fully described in the *Soil survey manual*

Consistence when dry. This attribute describes the rigid or brittle resistance of an 'air dry' aggregate to fragmentation or pulverization in the hand. The definition applies particularly to torrid conditions; in England a truly 'air dry' condition in a field soil must be considered as exceptional. There are six degrees for this form of consistence.

Loose
Soft
Slightly hard
Hard
Very hard
Extremely hard
} The field diagnosis is fully described in the *Soil survey manual*

If soil particles are cemented together by some secondary chemical there are three degrees of *cementation*.

Weakly cemented,
Strongly cemented,
Indurated.

Further descriptive terms are sometimes used to describe characteristics in a more general way, but are not now listed in the *Soil survey manual* (United States Department of Agriculture 1951). The following table, however, is taken from the old edition (published 1937) and gives an idea of the meaning of such terms when they are used.

Name	
Brittle.	Dry soil breaks with clean sharp feature.
Cellular.	Pore spaces of regular size throughout soil mass.
Cemented.	Soil aggregates bound together by a cementing agent.
Cheesy.	Characteristic of a *wet* soil. Elastic, bends without fracture.
Coherent.	Compact. Very compacted.
Compact.	Dense and firm without cementation.
Firm.	Moderately hard; fragments can be broken between fingers.
Friable.	Easily broken, and reduced to crumb structure. (Good agricultural tilth.)
Hard.	Very difficult to crush between fingers only.
Impervious.	Very resistant to water or root penetration.
Loose.	Soil of small aggregates and maximum pore-space.
Mellow.	Porous mass, softer than friable, no tendency to pack. (Good agricultural tilth.)
Plastic.	Readily moulded.
Soft.	Readily crushed between fingers.
Sticky.	Adhesive when wet, cohesive when dry.
Tenacious.	Very like sticky but applies more to a cohesive character when wet.
Tight.	Compact, impervious and tenacious, usually plastic.
Tough.	Bores easily but very difficult to break.

Since the American method aims at a literal interpretation of the observed facts, it works well in the field and the surveyor finds but little difficulty in learning the technique.

In Australia it has been found convenient to devise yet another system for the classification of consistence based upon the degree of force required to cause strain in a ¾ in cubic block of soil of specified moisture content after manipulations between thumb and fingers for about two seconds. The following is an outline of the method used by Butler (1955).

Scale of force

Force 1. A very small, almost nil force. Sometimes it is convenient to employ force zero to indicate actual separation, as in loose sands.

Force 2. A small but significant force.

Force 3. A moderate force.

Force 4. A strong force, but conveniently within the power of thumb and fingers.

Force 5. A very strong force at and passing beyond the capability of thumb and fingers.

Kind of consistence

Crumbly. Breaks into a fairly uniform population of *structural aggregates*.

Brittle. *Fragmentary* if the material breaks into *fragments of odd shapes and sizes*, and *pulverescent* if the material breaks into *ultimate soil particles*.

Labile. The soil first breaks into pieces and then these pieces reform or coalesce into balls or rods which are of plastic consistence.

Plastic. The material bends and does not break.

Pulverescence and coalescence

When work is done on soil materials the result may be either to break them down into smaller and smaller pieces until finally the ultimate particles are separated, or alternatively the final result may be a coherent plastic mass. The concepts of pulverescence and coalescence refer to the changes in soil materials as a result of work done on them.

Pulverescence is the tendency of a soil material to break down to fragments or to its ultimate particles when the specified amount of work is done on it, i.e. using for 2 seconds just the force required to cause rupture.

Pulverescence 1. Less than 10 per cent of the material occurs in the form of irregular fragments and/or ultimate particles. Pulverescence zero is sometimes used if the material consists of 100 per cent natural aggregates. This is the ideal crumbly material.

Pulverescence 2. 10–30 per cent of the material occurs as fragments and particles.

Pulverescence 3. 30–60 per cent of the material occurs as fragments and particles.

Pulverescence 4. 60–90 per cent of the material occurs as fragments and particles.

Pulverescence 5. More than 90 per cent of the material occurs as fragments and particles. This approximates to the ideal brittle form.

Coalescence is the tendency of an unsmeared soil material to assume plastic properties when the specified amount of work is done on it, i.e. using for 2 seconds just the force required to cause rupture.

Coalescence 1. Less than 10 per cent of the material occurs in the form of plastic balls or rods after working.

Coalescence 2. 10–50 per cent of the material occurs in the form of plastic balls or rods after working.

Coalescence 3. 50–90 per cent of the material occurs in the forms of plastic balls or rods after working.

Coalescence 4. More than 90 per cent of the material occurs in the coalesced form after working, but some subdivision of the primary piece results from working.

Coalescence 5. All of the material occurs in the coalesced form after working and there is no subdivision of the primary piece, but some fracturing can be seen at the commencement of the operation.

The ideal plastic material would correspond to coalescence 6, and the criterion for this state is that there is no fracturing to be seen at the commencement of work being done on it.

Item VII. Organic matter: nature and distribution

The organic portion of the soil is a manifestation of the biological reactions involved in the soil-forming processes. By a careful study of the visible evidence of the soil humus profile, much information may be obtained about the soil's life history. Since the main portion of the humus of the soil profile is derived from the surface vegetation, either by leaf-fall or root decay, the nature of the leaf litter, grass mat, or vegetable debris should be examined and measured. The effect of 10 cm of oak litter upon the soil profile will be very different from that produced by 10 cm of spruce litter, and such differences should easily be detectable by an observer. The approximate age of accumulations of leaf litter may sometimes be determined by carefully folding back, layer by layer, the annual leaf-fall until the botanical structure of the leaves becomes vague and merged into the 'raw humus' layer of acid soils, or the 'mull' layer of the less acid and neutral soils. The phrase 'organic matter (nature and distribution)' in the profile questionnaire will suffice for most field work. Hesselman uses the term *Förna* for all that undecomposed leaf-fall and litter which has not broken down into darker and more compact 'vegetable mould' with the loss of its botanical structure. If the observer records, however, as 'litter' that accumulation of loose unrotted vegetable material which retains its position by gravity alone he will not be far from most of the generally accepted ideas.

There are four main types of 'humus' that can readily be recognized in the field:

(a) raw humus, acid humus, or mor;
(b) mild humus, neutral humus, or mull;
(c) 'intimate' humus of the soil mass;
(d) mechanically incorporated organic matter which will ultimately become 'intimate' humus.

Such divisions depend primarily upon the differences in the botanical and chemical nature of the vegation material and the manner of its decomposition and disposal.

Raw humus is characterized by its excessive accumulation (slow decomposition) and, frequently, by the presence of structural remains of plants. A certain degree of lamination may sometimes be observed in the horizons of soils of the slightly podzolized type, wherein the raw humus zone shows a clearly defined boundary to the A_1 horizon. Raw humus is characterized by an extremely low base content. The author cannot recall ever having observed raw humus with a pH value greater than about 4·5.

Mull, on the other hand, contains sufficient absorbed calcium to allow of a crumb or grain structure with a generally 'loose' or 'porous' constitution. The black organic material is usually more intimately mingled with the soil mineral particles to form indistinct and merging boundaries ($>$ 5 cm) between litter and mineral soil. The pH of such mull layers varies from about 4·5 in Brown Forest Soils up to about 7 or just over in Rendzinas.

Hesselman (1926), and Romell and Heiberg (1934), have evolved systems for the classification of Forest Humus types. These classifications, however, were designed for the forests of the cooler and more humid climates, and do not necessarily apply to those of the tropics. In fact in the tropical rain forest there may be little accumulation of litter at all. Microbial decomposition and the activity of earthworms, when present, and the ubiquitous termites, lead to a very rapid mineralization of all the plant debris.

Romell and Heiberg (1934), an outline of whose vocabulary is given below, make use of certain of Hesselman's names in their main groups: Hesselman's *Förmultningsskiht* or decomposition layer is designated as H. They continue to use the term *Mull* but change the term *Råhumus* or Raw Humus to *Duff*.

Mull. A porous, more or less friable humus layer of crumby or granular structure, with diffuse lower boundary, not, or only slightly, matted.

> (i) *Crumb mull*. A coarse-grained mull, inhabited by large earthworms, usually in large numbers. This is the classical prototype of the mull group. Content of organic matter usually around 10–20 per cent or even lower, rarely over 30 per cent. Rich herbaceous vegetation, including a spring flora of geophytes such as *Corydalis, Mercurialis, Anemone, Arum* (Eu.), *Dicentra, Dentaria, Hydrophyllum, Claytonia, Arisoema* (Am.), is characteristic. Litter of loose leaves, or at times practically none because of the rapid decomposition.
>
> (ii) *Grain mull*. Differs from the crumb mull by its finer granular structure and the absence of *large* earth-worms. Flora like the preceding, but mostly poorer.
>
> (iii) *Twin mull*. A complex type of humus, consisting of one upper stratum with the characters of matted detritus mull or root duff (see below), underlain by grain or sometimes crumb mull. Flora poorer than on the preceding types, but includes mull plants.
>
> (iv) *Detritus mull*. A finely granular mull, rich in organic matter (usually over 50 per cent), looking like black sawdust. Flora variable, but always including mull plants.

Duff. A layer of unincorporated humus, strongly matted or compacted, or both, distinctly delimited from the mineral soil, unless the latter has been blackened by the washing in of organic matter. Flora usually completely lacks typical mull plants.

(v) *Root duff.* F-layer poorly developed, usually practically absent. Humus of the H-layer finely granular, like detritus mull; when dry, it can practically all be shaken out from the dense root mat which holds it together. Essentially a hardwood type.

(vi) *Leaf duff.* Laminated F-layer of matted leaves, H-layer much like the preceding. A hardwood and hardwood–conifer type.

(vii) *Greasy duff.* F-layer usually relatively little developed, often more or less fibrous, H-layer thick (usually 0·1 m or more), compact, but usually not very tough, partly or entirely black, muck-like, with a greasy feel when wet, shrinks strongly upon drying.

(viii) *Fibrous duff.* F-layer well developed; entire humus layer fibrous, more or less tough, but usually not very compact, showing little shrinkage upon drying. The flora of the most typical forms includes Hylocomia or Ericaceae (particularly *Vaccinium*), or both.

Since then Bornebusch and Heiberg have produced yet another system for the classification of surface organic matter[1]. Their system (shown below) was discussed and accepted with certain minor reservations at a meeting of the Forest Soils Sub-Commission of the Third International Congress of Soil Science in Oxford, 1935, and is given in full in the *Transactions* (see Bornebusch and Heiberg 1936). It was also accepted by the International Union of Forest Research Organizations.

I. The definitions of the kinds of forest humus must, in accordance with P. E. Muller, be based on morphological characters which can be easily observed directly in nature.

II. Two main kinds only are to be recognized: *mull* and *mor*.

III. *Mull: mixture of organic matter and mineral soil*, of crumbly or compact structure, with the transition to lower layers not sharp. Three forms are recognized:

(a) *Coarse mull.*[2] Coarse grain structure, organic matter very conspicuously mixed with mineral soil (usually 5–20 per cent organic content; in exceptional cases even considerably higher).

[1] By forest humus layers is understood the top layer of the soil, owing its characteristic features largely to its content of organic matter. This part is often described as A_0 and/or A_1.

[2] At the Congress in Nancy of the International Union of Forest Research Organizations in 1932, Section V adopted *true mull* for this form.

(b) *Fine mull*.[1] Fine grain structure. Organic content high (usually over 50 per cent).

(c) *Firm mull*. Dense compact structure, usually low content of organic matter, often less than 5 per cent.

IV. *Mor*. Organic matter practically unmixed with mineral soil, usually more or less matted or compacted. *Transition to mineral soil always distinct*. Often composed of two layers named (after Hesselman),[2] F-layer, i.e. fermentation layer, resting on H-layer, i.e. humified layer.

The F-layer consists of more or less decomposed litter, *still recognizable* and with rather loose structure.

The H-layer consists principally of finely divided organic matter *mostly unrecognizable* as to origin.

Three kinds of mor are recognized:

(a) *Granular mor*. H-layer[3] pronounced and fine granular in structure; lower part somewhat compacted. When dry very easily broken into fine powder when pressed by hand.

(b) *Greasy mor*. F-layer usually relatively little developed, often more or less fibrous. H-layer[4] thick, compact, with a distinct greasy feel when wet, hard and brittle when dry.

(c) *Fibrous mor*. F-layer well developed. Both F- and H-layers[5] fibrous but not compact. Many plant remains visible also in H-layer.

Peat

The classification of surface organic matter is not entirely complete without some reference to the various forms of peat. A simple system of classification has been drawn up by Fraser (1933), and depends mainly upon the appearance and behaviour of the material during handling. Fraser defines three main types:

(a) *Pseudo-fibrous peat*

Soft and plastic, rigidity and tenacity lost, fibrous in appearance only.

(i) *Cheesy peat*

Rigidity partly maintained under intermittent aeration.

---

[1] Section V adopted *superficial mull* for this form.
[2] The following definitions for F- and H-layers are not exactly in accordance with Hesselman (1926).
[3] H-layer described as *fine humus*.
[4] H-layer described as *greasy humus*.
[5] H-layer described as *fibrous humus*.

(b) *Fibrous peat*
Tough and flexible composed of scarcely altered remains of plants.

(c) *Amorphous peat*
Showing no recognizable plant tissues.

(a) *Pseudo-fibrous peat—characteristic of* Scirpus *moor.* Pseudo-fibrous *Scirpus* peat is a structural peat, being composed of recognizable remains of *Scirpus*, *Sphagnum*, and other plants. Of these remains the stems and roots are easily visible, so that the peat has a fibrous appearance. It is, however, quite plastic since the apparently fibrous structures have undergone fundamental changes so that their strength and tenacity are completely lost. The organic matter as a whole has so altered that the peat is capable of absorbing a large quantity of water and swelling considerably and, on the other hand, shrinking to a remarkable degree when slowly dried. This change in the plant remains takes place only in the complete absence of air, and it is important to note that if pseudo-fibrous peat is exposed to the air, i.e. if aeration is established, the fibres regain their strength and the peat becomes fibrous, while the amorphous matrix shrinks into black grains or brown encrustations upon the fibres.

(b) *Fibrous peat—characteristic of* Calluna *moor.* The nature of fibrous peat has already been indicated in the above section. Fibrous peat includes those structural peats in which the *strength and tenacity of the original plant tissues are retained*, and for this reason fibrous peat shrinks to a much less extent than pseudo-fibrous peat. It is not markedly different in appearance from the turf below which it occurs.

(c) *Amorphous peat—characteristic of* Molinia *moor.* By amorphous peat is meant that in which the *processes of decay have gone so far that a form of true humus or mould has been produced* from the peat-forming plant remains. In this kind of peat the remnants of plant structures are not any more visible than in the organic matter of an ordinary soil. Amorphous peat is dark brown or black in colour, and is composed of small particles indistinguishable from very humose clay soil, but of course it contains few or no mineral particles. The amorphous peat of *Molinia* moor varies from a black mud-like peat mass, which on drying becomes granular in appearance, to a brown-coloured peat of spongy texture which in its dry or wet condition is similar in appearance to *well rotted farmyard manure*.

Intimate humus

This material is found truly incorporated in the soil mass and varies in colour from black—through the browns—to greys. Intimate humus is not necessarily confined to the A horizon but can be found dispersed and recoagulated in the B horizon of Podzols, Vleis, and certain other ground-water soils. Highly decomposed intimate humus is frequently to be found in certain of the red and yellow tropical loams with black or grey surfaces. Certain other tropical soils show no visible humus whatever, neither do they effervesce with peroxide in the field, though they may contain organic matter by carbon estimation in the laboratory. This intimate humus has more effect upon the structure and constitution of soils than either of the preceding types, and particular care should be taken in its recognition and delimitation. Since it is so completely united into the soil mass, hydrogen peroxide and its visible characteristics are the only criteria for its description.

The description of the incorporated soil organic matter commences at the A_1 zone: its disposal around the soil particles, in interstices or in the body of the structural elements, should be noted. The degree and nature of cementation should also be determined by the sense of touch. Differentiation between cementation by humic or mineral material can usually be achieved in the field by means of acid, hydrogen peroxide, or a pocket lens. The extent to which humus is dispersed, mobilized, or transported down the profile is generally fairly easy to see. Calcareous soils tend to fix or immobilize humus near the surface, whereas leached and unsaturated soils tend to disperse and mobilize the humus down the profile into a secondary horizon. Sometimes basic ground-waters are responsible for the very black stain observable in Fen, Vlei, and Alkaline Peat soils. The colouring and mobility of humus in saline and alkali soils depends much more upon the moisture, sodium content, and the permeability of the soil mass than upon the actual quantities of humus present.

To ascertain the penetration of humus down the profile it is often convenient to pour a little peroxide down the freshly cut face, when effervescence ceases at the limit of intimate humus penetration. However, commercial peroxide usually contains acid and so will react with carbonates. This method may also prove erroneous in the field in the presence of concretionary (i.e. pedological) manganese dioxide, unless the precaution is taken to differentiate between the two. Concretionary manganese dioxide may only be crushed with a certain degree of force; elementary carbon is usually softer. Manganese dioxide usually reacts more violently with peroxide than does humus.

Mechanically incorporated organic matter

This refers to unhumified or only partially humified material *within* the soil mass, brought into any horizon by the decay of roots, the downwash of surface organic matter into old root or animal channels, or by the mechanical mixing produced by cultivation or faunal activity. This material exerts an important influence upon the depth of the development of the biological profile since it becomes trapped in constitutional fissures of the soil fabric ultimately to be converted to true humus and causing melanization of that portion of the soil mass in which it rests.

Item VIII. Roots

Observation of root development gives clues to differences in soil constitution, moisture conditions, and limits of aeration. For example, when tree roots cease to grow downwards, bend at a sharp angle, and run laterally there is an obvious reason, which it will behove the observer to find out and explain. The condition or nature of the roots, as to whether they are anchor roots, fibrous, old, young, dead, etc., and their quantities and distribution are all important items of general value and as such should be recorded. In turfy soils the nature of the sod layer is very important, the development of the grass roots being obviously correlated with the quality, population density, etc., of the vegetation cover.

Roots may be classified under seven headings:

Names of species of plant present if possible.

Quantity. Abundant—more than 100 per ft^2 of profile face.

Frequent—100–20 per ft^2 of profile face.

Few—20–4 per ft^2 of profile face.

Rare—3–1 per ft^2 of profile face.

Size (diam). Large— $> \frac{1}{2}$ in.

Medium—$\frac{1}{2}$–$\frac{1}{8}$ in.

Small—$\frac{1}{8}$–$\frac{1}{32}$ in.

Fine— $< \frac{1}{32}$ in.

The qualifying adjectives long, medium, and short may also be used.

Shape. Free-growing, distorted, etc.

Nature. Woody, fleshy, fibrous, or rhizomatous.

Health. Dead, alive, strong, or weak.

Age. Old, young, or of past or present vegetational cover.

Sometimes a soil horizon may contain characteristic layers of mycelia or colonies of mycorrhiza. This should be recorded.

In addition, the general description of the soil profile will be greatly improved if some graphic representation of the concentration or distribution of roots can be made in the profile sketch.

The significance of root-channels is dealt with in Item VI on soil constitution, and Item IX below.

Item IX. Water conditions

This item includes not only the movement of water through the whole soil mass but also the soil moisture as it moves or is held in the soil pore spaces.

Drainage of the soil mass. Drainage of the soil mass depends upon several local factors, such as

local climate,

mass and character of the vegetation,

physical nature of the underlying rocks,

proximity of sub-surface zones of permanent or intermittent saturation,

the texture and constitution of the soil mass itself.

Since the air held in the pores of a soil depends upon the amount of pore space not filled with water, it follows that as water increases, air decreases. As a general rule we may assume that a soil, the pore spaces of which are completely full of water, will be anaerobic. When all the water has left a soil then the pores will be full of air and the soil will be fully aerobic. Between these limits there are all possible combinations. Fortunately, there are well-defined field characteristics by which many of these combinations may be diagnosed and assessed. An excessive moisture-to-air ratio results in the phenomenon of *gleying*, and characteristic colours, mottling, and secondary chemicals betray the zone in which such conditions occur. We may assume, therefore, that if no gleying appears anywhere in the whole soil mass there is an adequate supply of oxygen for most of the year, and probably more air than water in the soil pores. On such assumptions it is possible to describe the *quality of drainage* in four *drainage classes.*

Excessive drainage. This implies that water moves rapidly through the soil mass and insufficient of it is retained for the normal growth of any species except those in true ecological equilibrium with it. This is somewhat difficult to assess in the field except by negative evidence such as the absence of diagnostic characteristics which would compel it to be placed in another drainage class. It is judged mainly upon the texture, constitution, and vegetation.

Temporary waterlogging of the surface by storm water may occur.

Perfect drainage (often called *free drainage*). This class represents the ideal conditions for general cultivated or managed crop plants. It implies that water moves easily through the soil to give ideal aeration, while at the same time the pores retain sufficient water for normal plant growth. It is diagnosed in the field by a fairly evenly distributed moisture content, and by the almost complete absence of any mottling, secondary chemicals, or reduction colours (gleying), while iron compounds in the soil mass are in a fully oxidized condition. Texture, constitution, and root development all assist in the diagnosis. Temporary waterlogging may occur from storm water but it very quickly disappears by percolation.

Imperfect or *poor drainage.* There are many facets to this class, depending upon the degree of development of gley colours and secondary chemicals. It implies that there is some fluctuation between aerobic and anaerobic condition at different times of the year. It is believed that under anaerobic conditions, probably in the presence of organisms or organic matter, the iron oxides become reduced and rendered soluble, when they may move up or down the profile to some limited extent. As the water level falls, air enters and reoxidizes these iron compounds to fix them as ochres in different degrees of hydration down to the limit of aeration.

For example, a 'very poorly drained' soil has this zone of oxidation very much nearer to the surface than does a soil that is merely 'poorly drained', while the zone of anaerobic colourings may go right down to the zone of permanent saturation or to the parent material as the case may be. Again, a very poorly drained or water-logged soil might not exhibit any ochreous mottles at all, but be greyish or greenish right up to the surface organic matter (peat).

Another important diagnostic feature is secondary manganese dioxide either as mustardseed-like grains or as dendritic patterns on stones in what may be termed the zone of 'maximum fluctuation' (see Gley, p. 38, and Secondary deposits, Item X below).

Impeded drainage. This class implies some definite obstacle to the downward percolation of water and is frequently associated with pans, rock pavements or other impermeable strata, perched water-tables, or a permanently high zone of saturation. It implies also that conditions at that depth are anaerobic, i.e. the zone of impeded drainage builds up a waterlogged environment which cannot drain away vertically, though lateral flow may occur. This lateral flow is responsible for the slow removal of soluble iron compounds from the anaerobic environment. When exposed to air these give the characteristic rusty scum so frequently to be found floating in ditches. Owing to the loss of this iron the gley horizon persists for a very long time even after the area is drained or the impedance removed.

Some observations on drainage

In a profile with impeded drainage there will always be paler patches in the mineral mass due to reduction and ground-water leaching; mottling may occur, but is not an essential criterion, since weathering of parent rock often produces a similar effect. In certain soils, the water-tables of which have been modified by artificial drainage operations, evidence may be found of two gley horizons. The old or upper gleyed zone will show signs of more recent oxidation or aeration in that brighter colours will be found on the edges and faces of the structural elements, the characteristic blue-green varnish will have disappeared, and the faces of the prismatic structure will have become modified towards the cubic type.

Well-drained soils are well oxidized to full colours. The size and shape of the structural elements, the soil constitution, and the true texture all exert their specific influences. Another important diagnostic factor is to be observed in the colour of the soil in the immediate vicinity of roots. When the channels of the living roots are picked out in lighter colours than the surrounding soil mass (i.e. a brown soil has its roots outlined in grey or green) it may be taken as a criterion of impeded aeration. The roots needing oxygen must take it from somewhere, and the fully oxidized iron compounds present in the soil mass become reduced. When a root dies, as it probably will in these circumstances, it will shrink and decompose by carbonization to leave a root channel through which air may pass. This air oxidizes the grey-green reduced material back to its full ochreous colours which outline the channels in rusty yellows and browns.

Water relationship and vegetation

During a soil survey in north Shropshire Davies and Owen (1934) made use of six general descriptions for the classification of their soils upon drainage and vegetation. The interdependence of these two items is extremely important. The following is a résumé of their classes.

Group I. *Soils developed under conditions of free drainage*
Such soils are Podzols and Brown Earths. These are then subdivided into

(a) *soils under natural vegetation,*
(b) *soils under cultivated vegetation.*

Sub-group I(a) is characterized by podzols carrying *Erica, Vaccinium,* etc. There is some doubt as to whether this podzolic condition is primary or developed as a result of the removal of the primeval forest.

Sub-group I(b) is characterized by a soil development that may be described as a Brown Earth (uniform SiO_2/Al_2O_3 ratio throughout profile). It has resulted from the enclosure and cultivation of the natural heath podzol to which it would rapidly revert if the land were to drop out of cultivation.

Group II. *Soils developed under conditions of impeded drainage*
Such soils belong to the Group of Meadow Soils.

Sub-group II(a) *Natural soils with high water-table.* Such soils are almost continuously submerged and bear a flora of aquatic plants resulting in the formation of fen peat (developed under anaerobic and neutral conditions).

Sub-group II(b) *Drained soils with a natural high water-table.* These are dominantly organic soils. Their depth depends upon the length of time during which they have accumulated and the nature of the vegetation. The subsoil is usually bleached, probably due to anaerobic conditions and reduction of iron compounds.

Sub-group II(c) *Natural soils with impedance due to impervious layers.* The primeval associations of these soils were probably wet woodland with a semi-aquatic ground flora such as *Aira, Juncus, Carex,* etc., or wet grassland. Since most of the areas once under these associations are now drained they have been converted into soil types of the next sub-group.

Sub-group II(d) *Drained soils with impedance due to impermeable layers.* The effect that artificial drainage has had upon the older associations is incalculable, since some of the low-lying land now under grass was probably dominated at one time by *Scirpus, Eriophorum, Carex,* etc.

Water relationship and soil texture classes

Glentworth (1954) has shown that there is a very practical side of this problem and the following text is taken directly from his paper:

'The amount of water a soil will hold against the pull of gravity depends on soil texture and may be referred to as inches of water per foot of the profile. It has been estimated that the amount of water available to plants in a naturally freely drained soil, for the different textural classes, is as follows:

Sand and loamy sands	$\frac{1}{4}$ to $\frac{1}{2}$ in per ft.
Sandy loams	1 in per ft.
Fine sandy loams	$1\frac{1}{2}$ to $1\frac{3}{4}$ in per ft.
Loams	2 in per ft.
Clay loams	3 in per ft.
Clays	$3\frac{1}{2}$ in per ft.

Assuming that the roots of agricultural crops reach 3 ft then the amount of water retained in a 3-ft column of soil varies from less than 1 in to almost 12 in. An eight-quarter oat crop will remove over 12 in of water, so that in regions of light textured soils it may be seen that even under our relatively humid climate, conditions of water shortage may seasonally affect the crops adversely.'

Soil moisture distribution in the profile. It is desirable that the relative wetness of the individual layers of the profile should be recorded.

For most purposes the terms *dry, moist,* and *wet* should be used in a comparative sense. A soil might be described, for example, as exhibiting a moist upper zone, a drier middle zone, and a wet lower zone, though of course the nature of the weather immediately preceding the examination must be borne in mind while taking notes. The degrees of wetness may be detected by eye and hand, and may very simply be classified:

Dry: structural elements are *visibly* dry and do not further change their colours when exposed to dry air but become darker when moisture is added.

Moist: structural elements *feel* moist and if exposed to further wetting will change their colours. Most soils will be able to be moulded into rods not more than ⅛ in diam. and ½ in long.

Wet: there is visible fluid water between structural elements or water may be squeezed from them. Their colours do not alter on further wetting.

Item X. Secondary deposits (chemical deposits of pedological origin)

Most chemical deposits are due to reactions in the soil solution. They are recognized and classified in the field according to their form and colour.

Efflorescences. These occur on the outer edges and rough surfaces of structural elements on drying out. They are finely divided, sometimes resembling a dusty powder and occasionally occurring as a bunch of very fine hairs or bloom.

Dendrites. This term is used when efflorescences on the faces of structural elements or upon stones give the impression of a picture of a tree or branching plant. Such designs are frequently produced by manganese dioxide in certain Brown Forest Soils.

Crusts. These represent a greater development of the efflorescences.

Veins and tubes. These are usually well defined as the fillings of old root channels and are frequently to be found in heavy soil under either woodland or meadow conditions. Petrified, fixed, or filled animal burrows and worm-holes may sometimes be included.

Concretions. Concretions usually occur as extensive zones of grain-like and nodular accumulations and are especially noticeable in the *murram* of the tropics or in the lower zones of the Black Earths of Russia and America, and in the Black Cotton Soils of Africa.

Streaks and interlayers. Streaks and interlayers usually occur when the whole of a zone is undergoing induration by the deposition of some cementing agent.

Humus columns. These occur in clays where, during drought, the soil mass cracks into deep clefts. When the rains commence topsoil falls, or is washed, into the cleft so formed. The subsequent expansion of the clay then seals a column of topsoil into the soil mass, sometimes to a considerable depth.

Chemical compounds and the common forms in which they may be found.

Crystals occurring as efflorescences, crusts, interlayers, and pockets. Carbonates, chlorides, and sulphates of the alkalis, being easily soluble in water, are readily carried in the soil solution to be deposited in the order of their solubilities, least soluble first.

Carbonates, chlorides, and sulphates of the alkaline earths obey the same general rules of distribution but in the case of the carbonates the transporting power of the transpiration current is governed by fluctuations in the carbon dioxide content. It is for this reason that accumulations of calcium carbonate are found much more widely distributed in different regional soil profiles than are other more soluble salts.

Pedological gypsum may occur in the form of twin crystals or 'earth hearts', but it more frequently occurs as a very fine powdery efflorescence on the faces of structural elements, or as pseudomycelia within the matrix of a structural element. It may be distinguished from calcium carbonate by its reaction with barium rhodizonate.

Amorphous powders. In certain soils white powdery deposits of dehydrated silicic acid may be observed as 'silica flour' which may be distinguished from other white amorphous substances by its insolubility and tastelessness on the tongue.

Irreversibly precipitated concretions and deposits. Compounds of iron, aluminium, managanese, titanium and phosphorus occur in a great variety of colours and forms.

Iron and aluminium sesquioxides occur in the orstein layer in the B horizon of podzols. Ferruginous compounds frequently develop as black or brown beans or grains and occur in great variety in humid soils in the temperate and tropical zones. Spots of brown-yellow 'eyes' and brown 'dots' of iron and manganese oxides are common in heavy clay soils with gley horizons.

Other ferruginous deposits occur as brown laminae and brown ochreous or brown and crimson spots and patches. Crimson spots are often to be observed in tropical soils.

Red and yellow ochreous nodules, ochreous veins, pseudo-mycelia, and rusty patches often occur in root channels and in interstices of heavy clay soils.

Dark brown to purple dendrites of manganese-iron oxides are frequently found on the faces of structural elements and stones of the oxidized fringe of the gley zone.

Calcium carbonate as 'white eyes', nodules, or 'puppets' may be found in the lower horizons of certain *Szik* soils, Chernozems, and also in Brown Forest Soils on heavy clay in which surface run-off exceeds percolation. Concretions of calcium carbonate coated by a dull varnish of manganese compounds are common in certain of the black soils of the Sudan.

Dark grains of limonite and manganese dioxide and sometimes haematite may be recognized by their streak on rough porcelain.

Haematite	Red	Lateritic crusts
Limonite	Brown	Meadow soils
Red and yellow ochres		Meadow soils and gleys
Manganese	Black	Meadow soils and gleys

Item XI. The distribution of carbonates and pH in the profile

A little acid is poured down the profile over the 'fresh' face, and the points at which reaction commences or ceases are noted, as is also the intensity of the reaction. Carbonates may be present either as relics of weathered parent material (limestone or dolomitic rock fragments) or as new pedological deposits (such as the carbonates of soda, lime, iron, or magnesia), the size, shape, and quantity of which should be recorded under their respective headings.

Care should be taken to differentiate between concretionary or deposited material and relic parent material. Concretions will, on fracture, show some tendency to concentric formation and, though they may possess curious shapes in aggregation, the general form of the separate units is spheroidal. White fluffy or powdery patches and pseudo-mycelia of calcium carbonate are often to be found together with concretionary materials in certain soils, but whether one is merely a stage in the development of the other it is difficult to determine. Crystalline calcium carbonate as calcite or aragonite is frequently to be found in A–C and C horizons of certain of the red and brown limestone soils of the Rendzina or Terra Rossa types, due to the downwash of dissolved carbonate from the surface into the interstices of the lower horizons which are already saturated with the products of hydrolytic weathering. Subsequent dehydration produces either the powdery or the crystalline form of deposit, but the writer has never observed hard spheroidal concretions in these types of soil.

Relic carbonates will of course resemble the structure, texture, fracture, and chemical characteristics of the parent rock, and can usually be recognized even in a fairly fine state of division by means of a pocket lens. Failing this, however, it may usually be found that new calcium deposits are very much more easily soluble in dilute acid.

There is a considerable difference between the visible reactions obtained on treating the face of the profile pit with acid, when the soil is in excess of the acid, and the treating of a hand sample in a small dish, when acid is in excess of the soil. It must also be remembered that the more finely the carbonate is divided the more vigorous is the initial reaction to acid. Thus a layer containing a low percentage of highly comminuted or amorphous carbonate could, in certain circumstances, give the impression of quite a high concentration, while a coarsely comminuted hard relic carbonate (e.g. dolomite) might hardly appear to react at all. The testing of the profile face therefore should be used mainly for the delimitation of the carbonate distribution while the estimation of the quantity should be made on a hand specimen.

A rough idea of the quantity of carbonate present may be obtained by treating a palm-full of the soil in a basin with an excess of 10% hydrochloric acid and observing and listening to the degree of effervescence as in Table 8.

TABLE 8

% CaCo$_3$	Audible effects	Visible effects
< 0·1	None	None
0·5	Faintly audible increasing to slightly	None
1·0	Faintly audible increasing to moderate	Slight effervescence confined to individual grains. Just visible
2·0	Moderate to distinct, heard away from ear	Slightly more general effervescence visible at close inspection
5·0	Easily audible	Moderate effervescence. Bubbles to 3 mm. Easily visible
10·0	Easily audible	General strong effervescence. Ubiquitous bubbles to 7 mm. Easily visible

pH measurement

The value of this determination depends to a great extent upon its interpretation, and for field work a very rough idea, perhaps to the nearest 0·5 pH, is sufficient. It is valuable to know that one zone is more or less acid than another, and that there is some gradation in the profile, but the absolute values are less important than a range of comparative figures.

Many field methods of sufficient accuracy are available and the choice must rest upon personal inclination. It must be remembered, however, that the more sensitive the test the more treacherous it becomes in the field, owing to difficulties of cleanliness. Perspiring fingers, extraneous dust and, probably worst of all, the instability of indicators in glass-stoppered bottles, are difficulties for which the field-man must be prepared.

Item XII. Fauna of the soil profile

Of the three classes of fauna discussed earlier (p. 27) it is mainly the creatures of group (1) (the intra-soil fauna) and group (3) (the dual environment group) that affect the characteristics of the soil mass. Observations and records should be made of the species, population, distribution, and effects of any creatures found in the soil and also—which is probably more important—the effects of the creatures that have at some time been there. Worms, mites, beetles, and all such small burrowing creatures leave behind them aeration channels while their secretions, cements, and coprolites produce characteristic effects upon the structure and constitution.

3. Soil Sampling

IN interpreting laboratory data a soil chemist is as much concerned with the local circumstances of the site as with the properties of the actual sample of soil received, so that some sort of site and profile description should always accompany a soil sample. The site and profile description should be made as full as possible.

There are two main kinds of sample. The primary samples are, of course, profile samples representative of the general pedological processes of the soil series. Secondly, a sample is collected from the arable layer of the soil, which may vary somewhat from field to field through cultivation and various systems of manuring and utilization though still belonging to the same soil series.

The profile sample

The taking of a soil monolith for a permanent record of the soil character is in the end the most satisfactory method to adopt, although if this is impracticable samples may be collected in cartons, as below. The advantages of the monolith sample often outweigh the disadvantages of the labour involved in its collection, by the preservation of the structure and constitution; at the same time, too, it is often easier to pick out horizonal differences when the sample is laid out on a well-lit bench than when the observer is cramped up against the side of a hole often under variegated light filtering through vegetation.

If it is also to provide samples for analysis a 5-in monolith is taken instead of the usual 4-in exhibition specimen. The telescopic lid (see below) is fitted at its maximum capacity for conveyance to the laboratory. The monolith is then faced to the 4-in (exhibition) limit by the careful picking out of the structural elements so that fair samples of each horizon are obtained. These are weighed and stones and fine earth are determined in the ordinary manner.

The monolith container is made of steel, and is 26 in long. Profiles to any depth may be taken, however, as containers are made to dovetail into each other. Sections may be taken separately in the field, and complete profiles built up from these in the laboratory without any serious disturbance of the natural soil structure. The advantage of the short sectional container is that each section, when full, is approximately a one-man load. The following is a description of the equipment and the method of using it.

A complete unit consists of container, lid, ends, envelope, slice, and bottom clip, the specification of each of which is given below.

Container. An open-ended trough, 26 × 8 × 4 in, of 20-gauge mild steel, accurately bent so that it is $\frac{1}{8}$ in wider at one end than the other (see Fig. 7). Holes for nails are accurately spaced near the ends to fix the

Plan of container

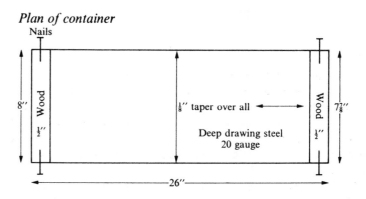

Side elevation of container and lid

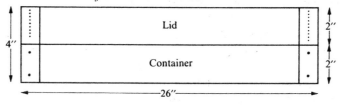

FIG. 7

wooden end-pieces and to coincide with holes in the lid. The taper allows for telescoping when composite profiles are built up from two or more containers.

Lid. The lid is made of the same material as the container. It is $\frac{1}{8}$ in wider at each end than the container, over which it fits closely. The sides are 2 in deep and have seven closely spaced nail holes at each end. This arrangement enables monoliths to be taken up to 1 in thicker than the 4 in of the container.

Ends. These are of straight-grained soft wood to fit the two ends of the container, and 5 in deep.

Envelope. This is no more than a larger lid of heavier gauge, made to fit flush over the container in order to take the extra inch of soil not obtainable by the 4-in container. It is not required if no extra soil is to be taken.

Slice. A piece of strong sheet steel or a flat spade blade to sever the soil at the base of the monolith.

Bottom clip. A piece of folded steel, shaped to fit over the lower end of the container and temporarily secured by two nails pushed through holes in the sides, and into the soil in the container. This clip prevents the sample from slipping out of the container during the process of lifting from the profile pit.

Method of collection

The operation requires two men. A site being chosen, the hole is dug and faced up ready for the container. The container (or the container in its envelope) is then pressed firmly against the profile with its wider and upward end 1 in above the soil surface. Two methods of cutting back may now be used according to the texture and consistency of the soil. With stony, light, and friable soils cuts are made as in Fig. 8.

The first cuts A–A are made with a 12-in knife or a coarse-set pad saw reversed in its handle to give a draw cut (the latter is the more efficient). A steady pressure is applied to the container so that it inserts itself into the cuts, but it is rarely possible to push the container right home. Cuts B–B (Fig. 9) are then made by chopping back with an adze some 3 or 4 inches outside the cuts A–A. Pressure on the container will now allow of its being pushed right home, and when completely full the cuts B–B are worked across to the point C. This is necessary to allow for the fall of loose material and the removal of such obstructions as stones or roots. In the case of soils which are likely to crumble or fracture the method of H. Greene (Sudan) has been proved to be of use. A 4-in bandage is wrapped round and round the container as the cutting back proceeds, so that the whole monolith is locked up in one solid piece.

When these cuts have been completed, the container is pushed half an inch downwards and a steel slice (or the base clip) is pushed under its base. If, however, the container is bandaged, this is impossible and must be done when the end-pieces are fitted. The man standing on the top of the pit then tilts the whole block of soil into the arms of the other man waiting in the pit. The block of soil in the container now resembles Fig. 10. The face of the monolith is cleaned off

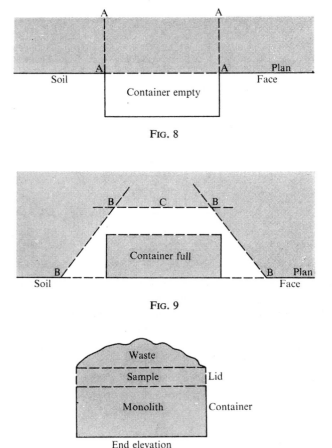

FIG. 8

FIG. 9

FIG. 10

flush to the edge. The lid is pushed home over the container and the end-pieces fitted and tacked into place, any loose spaces being carefully plugged with cotton-wool.

For stoneless clays and loams, which will cut clean, a simpler method may be used as shown in Fig. 11.

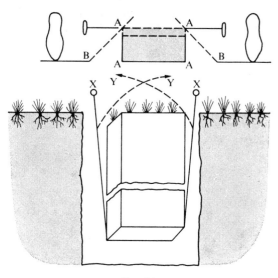

FIG. 11

The initial cuts, A–A and B–B, are similar, but instead of cutting back to an apex at C a cheese wire is pushed under the base clip by a small prong or by a screwdriver with a notch cut in the blade, and an operator standing astride the monolith draws upward with a sawing motion as in X–X. With two men drawing it is better if the wires are crossed as in Y–Y, which is in general the easier way.

For samples containing many roots or soft stones a section of a band saw set in two handles, bent round the whole container and used by the man in the pit with a sawing motion from side to side, has proved very useful.

Packing and transport

The sealed monolith unit weights between 60 and 70 lb. This is a full one-man load and cannot normally be carried far before some other form of transport is needed. Single units will travel safely for very long journeys when laid on sacks on the floor of a van. In rough country a properly plugged unit may safely be carried slung on a pole between two bearers. For transport by rail or sea, units travel best when packed in fours in a strong wooden box. The box should be at least 2 inches larger all round than the contained units and the spaces packed with rammed straw, particular care being taken with the corners. When so packed and transported, monoliths have arrived in Oxford in perfect condition from all over the world.

On arrival at Oxford, the soil monoliths were laid flat on a bench and the lid removed to allow the sample to dry and 'structure up' naturally. During drying the plugging is carefully removed and the face is constantly picked by hand and sucked by a vacuum cleaner to remove disintegrated fragments and dust from the constitutional fissures. The monoliths may remain as permanent exhibits in this condition, but they are bulky and difficult to store and handle. Therefore, there are many advantages in reducing the dimensions and mounting the specimens in such a way that they may be stored in library racks when not required for demonstration.

The preparation and preservation of soil monoliths of thin section[1]

Cutting down and mounting

A piece of five-ply board, 6 in wide and of the required length (normally units of 24 in), is prepared and *around* this is fitted a frame (0·5 × 1·5 in) to form a tray of an overall width of 7 in and an internal depth of 1·0 in (see Fig. 12 (b)). The frame is not fixed to the base board. The monolith that is to be mounted is laid down flat with the open face upward. All superfluous knobs and excrescences of the structural features on the exposed face are gently rasped down to give a smooth surface, care being taken not to interfere with constitutional cracks. Roots are carefully cut off flush with the face with side-cut pliers. Protruding stones, if of a soft material, may be sawn or filed off without removal, but hard stones must be eased out, chipped to a flat face, reinserted and fixed with clear Bostik. The smooth face is freed of dust with a vacuum cleaner and painted with two or three applications of a solution of 3 % cellulose acetate in acetone until the top 0·1 inch is well impregnated. While the surface is still wet a thin layer of cotton wool slightly larger than the dimensions of the face is laid evenly overall and pressed down. The protruding edges of the wool are then folded upward to form a tray 6 in wide and to leave a margin

[1] See Clarke 1962.

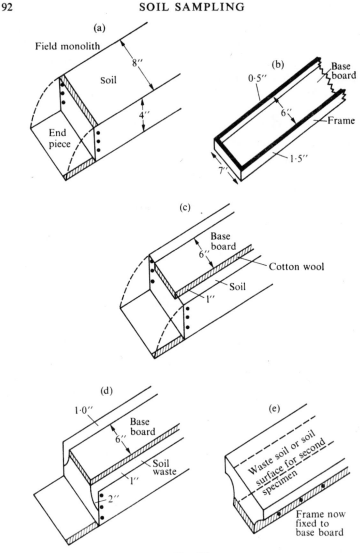

FIG. 12

of bare soil 1 in wide on each long side (Fig. 12(c)). Into this
tray is poured a solution of 10% cellulose acetate in acetone which
should not overflow. This thick solution must be spread evenly
with a flat brush, and left long enough for it to penetrate through
the wool and reach the previously treated soil face beneath. The

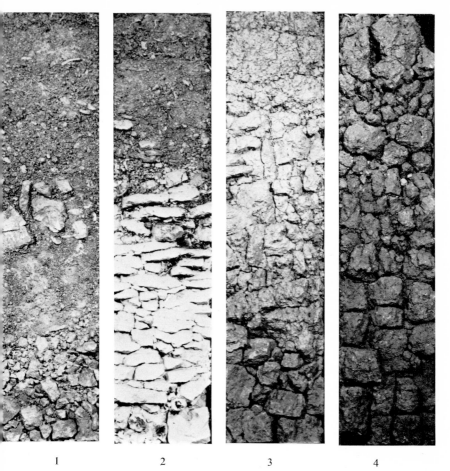

1   2   3   4

1. Limestone, Cotswolds.   2. Chalk, Chilterns.   3. Wealden Clay, Kent.
4. Upper Greensand, Berks.

SOIL MONOLITHS ILLUSTRATING STRUCTURE
AND CONSTITUTION

PLATE 1

(Soil Museum, Department of Agricultural Science, University of Oxford)

5	6	7	8

5. Oxford Clay, Oxon. 6. Middle Lias, Banbury. 7. Old Red Sandstone, Brecon (Hill peat). 8. Old Red Sandstone, Brecon (Peaty podsol and thin iron pan).

SOIL MONOLITHS ILLUSTRATING STRUCTURE AND CONSTITUTION

PLATE 2

(Soil Museum, Department of Agricultural Science, University of Oxford)

base board is now laid on the sticky wool leaving a margin of 1 in between the board and the edges of the steel trough. The wooden end-pieces previously loosened are now folded down out of the way (Fig. 12(c)), the base board pressed home, held down by weights, and left to set for 2 days. By this time soil and board should be securely joined and the monolith may be cut down to fit the base board. Cuts with a large knife about 2 in deep are made along the sides of the base board between it and the steel sides of the trough. The marginal soil is then scraped out and discarded (Fig. 12(d)).

Mounting

The assembled frame is now fitted over the base board and pushed downward until it fits snugly to the top edge of the base board where it is secured by screws. The whole block is inverted and the steel trough is gently raised to leave the soil mass free at the top but securely fixed on the base board now beneath (Fig. 12(e)). The loose debris is removed with the minimum of shattering, either by hand picking or with suitable tools, to leave a roughly structured ashlar some 1 or 2 in thick. Protruding stones are dealt with in the manner already described. Alternatively, instead of rejecting the debris another base board may be stuck down as described and a duplicate thin monolith prepared by splitting the whole soil mass longitudinally.

Preparation of the face

From the surface of the rough ashlar structural units and loose fragments are removed until the desired thinness is obtained. Throughout this process the surface is exposed to continuous suction from the nozzle of a vacuum cleaner so that the structural characteristics are clearly defined and remain free from dust. A final cleaning is done by directing the nozzle at the *constitutional* cracks. The clean surface then requires to be fixed.

Fixing

The specimen is tilted at an angle of about 10° with the top horizon (A) at the lower end; a solution of 3% cellulose acetate in acetone is poured on the raised end while its flow is controlled with a flat brush. This control must be so exerted that particles are not dislodged and air bubbles are eliminated. The aim is to permeate the whole layer of soil so that the fixing solution reaches the cellulose acetate in the wool on the base board. Several applications may be needed before this is achieved: 'little and often' is the recipe. Sometimes stones may not be sufficiently fixed by the dilute solution. In such cases the specimen must be allowed to set and then with a fine brush a concentrated (10%) solution is made to infiltrate into the crevices. Alternatively, each loose piece is backed with Bostik and reinserted into its correct position. Some soils, particularly when there is a high content of roots, may give a white 'bloom' of cellulose acetate. This 'bloom' may be removed by brushing with a very dilute acetate solution or warm acetone (not over 50°C).

After the specimen has 'set' it is sufficiently strong to be erected in a vertical position without danger of collapse. (At the time of writing a 12-ft specimen is still standing safely after thirty-two years in the Soil Museum at Oxford.)

Hand specimens

It is sometimes desirable to preserve individual soil fragments, for example structural units, sod layers with their root systems, etc. To do this the specimen is first air dried and then oven dried at 50°C until it is at the same temperature throughout. While warm it is immersed in a bath containing a solution of 3% cellulose acetate in acetone until bubbles cease to appear, and then allowed to drain and dry on a sieve. Any white 'bloom' may be removed as previously described. Small specimens such as soil crumbs may be fixed by immersing them and sieving gently under the solution. During draining and drying the particles are kept in motion by teasing until they set individually. Specimens so fixed may be handled without breaking down.

Individual samples

Though many elaborate tools are available for the sampling of soils, the only tool upon which the field-man can absolutely rely is the spade. No boring tool yet devised will entirely eliminate compaction in some form or other, though certain implements of the gouge type, or cheese-sampler type, get very near to perfection. There are many of these on the market and there is but little to choose between them; they all work in certain soils, but none of them works in all soils.

Paper or linen bags for the collection of samples are not recommended; samples dry out and grind down, so that structural and constitutional features are destroyed. Bags also have to be of the very finest material to prevent some loss of the finer fractions. In fact, on receiving soils from long distances, it is very rare indeed to find the sample completely intact. If bags must be used polythene is the best material. Much of this trouble can be got over, however, by the use of ordinary paraffin-waxed cartons, the waxed tops of which, if run round with a lighted match, will melt sufficiently to seal the lids and so preserve the moisture.

Good sampling involves very careful differentiation of horizons, particularly when they are merging. A horizontal cut with the sampling tool (a butcher's knife) should first be made between the horizons to emphasize the sampling limits, and then two long and deep vertical cuts should be made in the face. If the zone shows little or no structure, a block of the soil may be cut out and neatly placed bottom side upward into a carton and securely packed. If good structural features are apparent, individual elements should be picked out and used as the representative sample of the zone. Friable soils or soils with loose constitution are usually taken more easily by lifting out the whole horizon with a spade and then dropping the whole mass cleanly into a carton so as to preserve as much of the zoning as possible. This procedure seems perhaps a little fussy to the spade-and-auger man, but when the samples are laid out on the bench in the laboratory the additional information obtained for this little extra amount of care is often of very great value. It is a somewhat moot point as to whether consecutive samples should be taken exactly contingent to each other or whether it is better to leave a small margin as waste. As a general rule the writer has found that a better horizonal differentiation may be obtained by rejecting as 'waste' all that material disturbed by the cuts between horizons. Particularly is this useful when sampling merging horizons, although it is a point upon which it is hardly wise to dogmatize.

Sampling of the utilization layer

This sample should be, of course, a composite one and representative of the immediate requirements of the land user. It should be taken from a sufficient number of points to eliminate local irregularities, though what this number should be is impossible to foretell: six per acre is a minimum. The depth to which the sample should be taken depends a great deal upon the crop involved; in the case of grassland the sod-layer should always be taken separately, and then a further sample should be taken to the limit of the next visible change, but should not normally exceed about 15 in. On arable land the layer actually turned by cultivation should be taken, with a further sample to the first visible change if desirable.

The best method by which to take the sample is to open a small hole about a spit or so deep with a spade and then take a thin but uniform slice from top to bottom, cutting it off at the required depth while laid out on the spade. Several such slices are taken, thoroughly mixed, and fairly sampled before leaving the field. Final mixing and sampling will of course be done during the preparation of the 'fine earth' in the laboratory. Sampling by trowel is only permissible if care is taken that a uniform sample is obtained from a *cylindrical* hole and not a *conical* one. Auger sampling is usually done with a tool of larger cross section than that used for mapping, but it is an unreliable practice and should not be done except to explore the deeper horizons below the limit to which a pit has been dug.

While sampling, it should always be borne in mind that the most meticulous analysis or laboratory investigation is rendered completely useless unless a fair sample has been obtained in the first instance. When it is realized that an acre of soil down to about 9 in deep may contain anything between two and three million pounds of soil, the taking of a representative sample of about 2 or 3 lb is a distinctly difficult practical problem. Very careful site selection is obviously essential.

4. Soil Survey and Mapping

WHEN the first edition of this book was published in 1936, the mapping of soils as an aid towards rational land use was in its infancy. By now there are few countries that have not undertaken some sort of organized census of their soils. Various systems of soil classification have been developed for this purpose. The placing on record, by means of a soil map and its memoir, of the information gathered from laboratory studies and from observations and interpretation in the field, is the natural corollary of all that is discussed in this book.

Use of maps

It can be assumed that the student has a certain amount of knowledge of field sketching and mapping, and that as a surveyor he has access to adequately contoured base maps of suitable scale

Map scale	R.F.	$\frac{1 \text{ inch}}{\text{on map}} = \frac{\text{yd on}}{\text{ground}}$		$\frac{100 \text{ yd on}}{\text{ground}} = \frac{\text{inch}}{\text{on map}}$	
50·688 in to 1 mile	1/1250	1 in =	35	100 =	2·8
25·00 in to 1 mile	1/2534	1 in =	70	100 =	1·4
6·00 in to 1 mile	1/10 560	1 in =	293	100 =	0·34
4·00 in to 1 mile	1/15 840	1 in =	440	100 =	0·23
2·534 in to 1 mile	1/25 000	1 in =	700	100 =	0·14
1·00 in to 1 mile	1/63 360	1 in =	1760	100 =	0·057
0·5 ($\frac{1}{2}$) in to 1 mile	1/126 720	0·1 in =	350		
0·25 ($\frac{1}{4}$) in to 1 mile	1/253 440	0·1 in =	700		
0·1 ($\frac{1}{10}$) in to 1 mile	1/633 600	0·1 in =	1760		

and quality. Nowadays, field mapping may also be done directly on to air-photo prints. The detail so obtained can be of a high degree of accuracy and its transference to a base map is a comparatively easy task.

In this country the maps of H.M. Ordnance Survey and H.M. Geological Survey supply practically all the detail that is needed. 1/25 000 maps (approx. 2½ in to 1 mile), contoured at a 25 ft vertical interval, show all field boundaries, woodlands, parks, etc. These are excellent for soil survey on the ground and for air-photo interpretation. For 'special problem' surveys the 1/2540 (25 in to 1 mile) may be used. Maps warp and stretch on service and so absolute accuracy is hardly to be expected. The list of scales on p. 97 has been drawn up to help the beginner to transpose from map to ground.

A comprehensive list of scales and equivalents is given on the back end-paper.

Indoor preparation

Before the surveyor goes out to the field at all he should spend some time in becoming familiar with the general conditions prevailing in his area. From studies of maps, air photographs, and relevant literature, etc., he should endeavour to construct a mental picture of the main and mezzo-topography, the solid and drift geology, and the distribution of land utilization (woodland, parkland, wastes, arable, etc.). He should also study the natural drainage (which is usually fairly obvious) with special reference to any artificial improvements, such as the locking of rivers and the distribution of canals, leats, *rines, dykes*, etc. He should next study the accessibility of the area and work out a scheme of traverses that will allow him to cover the greatest possible area of ground with the minimum amount of recrossing of already mapped areas. It is often useful to select a road, railway, or natural feature from which offsets may be made.

In unmapped country, or where adequate base maps are not available, much valuable information may be obtained from a study of aerial photographs. When examined under a stereoscope overlap photographs give a beautiful picture of the 'lie of the land', the drainage, and general utilization. The full interpretation of air photographs is only possible by the expert, but with a little practice a novice may become quite proficient in the recognition of localities and the differentiation of regional characteristics. Even in well-mapped country the use of air photographs for location finding, etc., greatly speeds up the work.

The field slip used by the Soil Survey of England and Wales is carried in a stiff leather case that is used as a drawing board. This was not always as convenient as it might be, especially when making use of air photographs, so the author used a small drawing-board of five-ply wood with the map or air photograph, which may be interchangeable, pinned beneath a sheet of stout *clear* plastic, where they are clearly visible. Plastic sheet can take and retain grease pencil markings. Being waterproof it protects the photograph or map from rain or dirt and may be washed clean on return to base. Mistakes may be erased by rubbing with the finger. Before using the plastic in the field the writer usually traced the drift geology and traverse lines on to it with coloured grease pencils and then used black for the soil mapping. To transfer data from the plastic to the fair map the plastic is fixed on a glass drawing board. The map is placed over this, and the whole securely fixed with Sellotape along the top edge. They are illuminated from below with a *cold* lamp. It is essential that the light is cold since the plastic and the paper are liable to expand and contract at differing rates while the grease pencil marks may run or smear on warming.

Field mapping

If the indoor work has been efficient and the area to be surveyed is covered by good topographic maps this part of the work is merely a matter of finding and delimiting the boundaries by sampling with an auger around the typical sites worked out beforehand and exemplified in the profile pits. It may require a certain amount of hard manual labour. The correct siting of profile pits is the whole basis of an efficient survey and cannot always be done beforehand.

In countries where the surveyor has to conduct a survey over ground still to be opened up (for example tropical bush) he may find himself compelled to add to his map much more matter of a purely topographical character. If air photos or air maps are not available in such cases, quite important details of aspect, altitude, slope, and outcrops are difficult to assess without an actual determination upon the site itself.

There are two general methods of attack that may be adopted, depending mainly upon whether the area is 'open', as in a grass or arable region, or 'difficult', as in woodland or in regions of broken topography.

Open country survey

The surveyor walks to a point selected to represent one of the 'sites' in his area. There he bores about with the auger until he is satisfied that he has found a profile representative of the site. Here he prepares his pit and marks his map with a triangle and number (for example ⚠). Having described the profile, he traverses towards a neighbouring 'site', taking borings either at regular intervals or in response to changes in vegetation or environment, until he observes a change of sufficient significance to warrant further investigation. Only the field-man can decide on this point and this is only possible after experience on the spot and cannot possibly be described here. It may help, however, if certain points are dealt with.

Changes in soil profile may occur in any of the following circumstances:

(1) a change in the nature of the lithological material,
(2) a change in topography,
(3) a change in the vegetation, which if on arable land may be looked for in the nature and quality of hedgerows, timber, etc.,
(4) changes in colour, of soil surface or 'termite mounds',
(5) a difference in the sound produced, or feel to the feet[1], when walking or stamping. This test is frequently of great value to the experienced field-man.

Much may be learned from the 'layout' of fields on air photos since in any country that has been intensively cultivated for a very long time, field boundaries 'tie in' with both soils and topography far more frequently than not.

If the surveyor travels from, say, permanent grassland with a good sod formation to old arable land low in humus, he must be prepared to discount the A horizon to some extent and work on the *depth* of horizons and the characteristics of the B, G, and C horizons. He must bear in mind that he is mapping soils and is not mapping systems of utilization or the idiosyncrasies of neighbouring farmers. For the bringing into agricultural use of the land of virgin prairie or steppe the Soil Series map is of great value, but in areas which have developed into a sort of large-scale pot culture the Series map may prove of little assistance to agricultural practice, because existing utilization differs from one area to another on an otherwise uniform soil.

[1] Or to the bottom of the surveyor sitting in a Land-Rover.

Soil survey in 'difficult' country

By 'difficult' country is meant areas under forest, or broken topography, etc., in regions where difficulty of access or traverse requires more than normal soil survey technique. It is in such areas that air photos really prove their value for locality finding, selection of routes, and the definition of relief. Feasibility of access, etc., will show more clearly on a good air photo than on the corresponding map. With or without air photos the art of locality finding must be learnt by experience, but the beginner may make a very good soil map with the aid of compass bearings, chaining, or pacing, and inspection pits at arbitrary intervals. By bracketing between inspection pits, soil boundaries may be sketched in with reasonable accuracy, but depending greatly, of course, upon the spaces left unchecked. In ordinary woodland survey in England the writer has found that if 'lines' are run at about two- or three-chain intervals (at right angles to the main and mezzo-topography whenever possible), with inspection pits at regular intervals and additional pits at every change in mezzo-relief, a very small area of land remains unclassified after a little bracketing. The great temptation for the beginner to use rides, fire avenues, and odd small clearings as sites because they 'look nice' must be discouraged, though they may, and should, be used as location points wherever possible. In the preparation of the inspection pits it is not usual to dig more than the absolute minimum necessary to give the surveyor the information he requires; the standard profile pit is used only for the full recording and sampling of each Soil Series. A good example of the practical application of these methods is shown in the recent work of H. Vine and his associates during their soil survey in western Nigeria (Vine, Weston, Montgomery, Smyth, and Moss 1954).

5. The Use of Air Photographs in Soil Survey

Air photograph reading and interpretation

The method of research is common sense refined to the point where it largely eliminates errors characteristic of common sense. (E. L. MERRIT)

REFERENCE has frequently been made to the work of Dokuchaiev as the foundation from which many systems of soil classification have since developed. It is remarkable therefore how very rarely air photo readers appear to have noted that the five points of Dokuchaiev's classification are nowhere so clearly apparent as on an air photo. It may be recalled that Dokuchaiev observed that 'every dry land vegetative soil is formed as the result of the functioning of five factors, viz:

(1) The nature of the parent material (mineral).
(2) The mass and character of the vegetation.
(3) The age of the site.
(4) The relief of the site.
(5) The climate of the locality.'

The first four are obviously capable of being photographed. Implicitly (2) covers all biological processes. The mass and character of the vegetation depends upon natural, semi-natural, and entirely artificially induced conditions. The major controlling influences are those arising from man's utilization of natural resources of which the results may readily be recognized on air photos by the utilization patterns. Truly natural sites may equally well be recognized from the lack of such patterns.

It is not intended here to deal with the problems of air photography or air survey, nor the principles of optics, because there are numerous excellent textbooks dealing with these subjects. An attempt will be made here to show how air photo interpreters or less highly trained air photo readers may learn something about soils.

It must be realized from the start that air photos directly show only the external or environmental features of soil. With experience so much more may be deduced from the study of these features that a very shrewd idea may often be obtained as to what is underneath as well. No intelligent photo reader would suggest that ground soil survey could ever become redundant; it is in fact the basis of the use of air photos, because unless a person knows what an object is on the ground he cannot recognize it from the air. For this reason it is easier to make a trained soil man into an efficient photo reader of soils than it would be to make a trained photo interpreter into an efficient soil surveyor without adequate field training in soil.

When a field observer has learned what a certain set of features represent in the field, he knows what to look for and recognize on an air photo. There are three degrees of mental effort to be made. First perception, secondly observation, and lastly interpretation.

The study of air photos for the recognition of soils is usually made without the use of elaborate photogrammetric apparatus. Single prints are first examined by eye to obtain an impression of the character and distribution of the various objects likely to help in the diagnosis. Later simple stereoscopic study of 'stereo pairs' of contact prints provide three-dimensional detail.

Prints should be of A quality (i.e. good definition of detail and free from cloud) obtained from panchromatic film developed to 1·2 gamma. Very simply, the higher the value of gamma, the greater is the contrast of a negative. It is generally advisable to photograph with a 'minus blue' or Wratten 55 (green) filter. The 'minus blue' filter cuts out haze while the Wratten filter cuts out the reds and blues and intensifies the greens. The focal length of the camera and the altitude of flight are arranged to give a contact print of the scale required,

$$\text{e.g. R.F.} = \frac{\text{Focal length in inches}}{\text{Height in ft} \times 12}.$$

Experience has shown that A quality prints at a scale of 1:10 000, which is approximately that of the 6 inches to 1 mile O.S. map used for field survey, give good enough definition for general soil purposes. For reconnaissance and for the construction of 'mosaics' smaller scales of 1/26 000 or less[1] are generally used. Contact prints on this

[1] The terms 'large' and 'small' scale often lead to some misunderstanding. On prints of the same size, the *smaller* the *R.F.* the *larger* is the *area* of ground.

scale may be examined under the reflecting stereoscope without magnification for broad environmental detail, and with the binocular magnification for local detail.

GENERAL METHODS

Prints

In Britain prints may be obtained from two main sources: (a) from commercial organizations that specialize in this work and supply the client with material exactly to specification: it is usually of the very highest quality but is unfortunately somewhat expensive; (b) from official sources, material that has mainly been obtained from operational sorties flown by the various armed forces. It is with the latter that we are more usually concerned in Britain. When an order for prints is made, the exact area should be specified by the proper grid references taken from the 1-in O.S. map. It is better still however, to send an annotated copy of the map itself since accidents sometimes do happen when referencing. The kind of 'cover' required should be clearly stated, particularly the scale, time of year, size, and number of contact prints. Sorties may have been flown in any direction; it often happens that they cross or overlap each other at any angle and give duplicate cover over odd areas. This is frequently a good thing since the sorties will amost certainly have been flown at different times and seasons which enables features to be compared under different conditions of growth, light, moisture, etc. When received, the prints should be accompanied by a plot of the sorties on a map. They are laid out consecutively and checked against the plot by examination of their 'titling strips'. The titling strip runs along one side of a print and gives the essential details of reference and certain additional information.

The titling strip of official British prints. The strip may lie along any edge of a print and is not indicative either of orientation or direction for viewing. It gives the following items in order.

Print number. This is a four-figure number, when marked by stamping, so that print number 6 in the sortie reads 0006 and so on. Sometimes prints are marked in handwriting, in which case the four-figure enumeration is not used, the prints simply being numbered consecutively.

The position of the camera. Numerous symbols are used to describe different cameras and their positions in the aircraft but for most general purposes two will suffice. A capital V as a single letter indicates one single vertical camera. If more than one is used the notation becomes V1 and V2, and they are numbered from nose to tail. A capital F as a single letter indicates a 'fan' of two or more cameras. The F is followed by two digits: the first indicates the number of cameras in the fan and the second the position of the camera in the fan. The cameras are numbered from port, e.g.

F21 is one of a pair of cameras and is the port one.
F22 is one of a pair of cameras and is the starboard one. so:

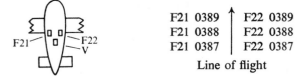

F21 0389   F22 0389
F21 0388   F22 0388
F21 0387   F22 0387

Line of flight

The *similarly numbered* prints overlap each other laterally by 20 per cent to give *split pairs* for the coverage of an area too broad to be taken in by one camera. They are *not* stereo pairs.

Stereo pairs of photos are taken by the *same* camera as it proceeds along its line of flight so that ground covered by one exposure is not out of range before another exposure is made. Thus F21 0387 will be covered by 60 per cent of F21 0388.

The squadron number and service flying the sortie.

Sortie number. Each squadron detachment starts its numbering at 0001 and continues to 9999 when it restarts at 0001.

Date. This is the date on which the photo was taken.

Time. As shown this is the *mean* time of the sortie and is given in G.M.T., indicated by a small z, e.g. 1400z is 2 p.m. G.M.T.

Focal length of camera in inches, e.g. 20 in.

Altitude. This is given in feet and is the *mean* height of the flight from O.D.

Security classification.

The following example is taken from the titling strip of a $7 \times 8\frac{1}{2}$ in contact print:

0211. F21. 58/RAF/1472. 24 June '54 = 1400z 20 in 16 666.
RESTRICTED.

The double hyphen separates 'references' from 'additional information'.

Line of flight and principal point. Midway along each edge of the print will be found special marks, arrow-heads or straight lines, projecting just into the picture. These are collimation or fiducial marks. Straight lines drawn to join opposite pairs of these will intersect at the *principal point* of the print. This point is the optical centre of the print and is the only spot on the photo where the object is vertically beneath the camera, is true to scale, and in true plan. The farther away an object lies from this point the greater will be the distortion. This is called 'radial displacement' and affects, among other things, perception of scale and height. Slopes running towards the principal point are exaggerated while those running away are diminished.

In order to obtain impressions of the height of topographic features, it is necessary to use a stereoscope. To make the proper use of this instrument the line of flight must first be found because stereo vision is only correct if made with the eyepieces either along or parallel to the line of flight, while the print is orientated correctly for shadow. To determine the line of flight the principal points of consecutive prints must first be found. Then each principal point must be transferred to every print on which it appears (e.g. a chimney is at the P.P. on one print but is nearer an edge on the next). These principal points are then joined up and extended to the ends of the prints, when the lines so obtained will show the line of flight. The line of flight does not always run in such a way that prints match to a perfect fit; they may overlap each other obliquely. Sometimes the plane may drift away from its set course, or at other times it may remain on course only by a crab-like motion. Either possibility results in a staggered array of prints when laid out on a table, and unless the line of flight is determined, stereo examination is made unnecessarily difficult for the beginner because a good deal of correct overlap is lost.

Mosaics

These are made to help the observer to visualize the 'lie of the land' over the whole of the area or any part of it, and to act as an index for the location of particular prints.

Photos are obtained from successive runs over the area with lateral overlap between the runs to produce 'block cover'. The best prints for mosaic-making are obtained by vertical cameras. The mosaic is made by laying out all the prints to give what is virtually a composite photograph of the whole area. The prints are then cut with a razor blade, just sufficiently deeply to go through the emulsion but not the backing, in a series of smooth curves. Each print is torn very gently to leave a face edge of razor-blade thinness so that, when pasted down on to the print beneath, every detail of tone and coincidence will run smoothly together without any ridges or false boundaries. The art of cutting and tearing is the elimination of displacement. When all this has been completed, the whole is photographed for reproduction at any desired scale. Such a photo is virtually an air map of the area.

It is not always necessary to make a 'stick-down' mosaic; much useful information may be obtained by taking alternately numbered prints (1, 3, 5, 7), placing them on the table to overlap with proper coincidence of detail, and then lightly connecting the run with transparent Sellotape. This economizes in prints, yet still permits stereo study of other alternates (2, 4, 6, 8, etc.) although under conditions of exaggerated eye base.

Stereo examination of prints by the pocket stereoscope

The pocket stereoscope is a simple instrument consisting of two small lenses with a magnification of about $\times 2\frac{1}{2}$, set in a small frame on legs. The lenses are set so that one may see a left-hand object through the left eyepiece and a right-hand object through the right eyepiece, yet when these objects are placed at the 'inter-ocular distance' of the observer in correct focus, their images merge to give a three-dimensional effect. The following drill is given here as the clearest and easiest way of learning to use a simple stereoscope.

Stereo vision. The viewer's inter-ocular separation must first be found. To do this take a sheet of paper and mark a dot upon it. Place the stereoscope over the paper so that the dot appears immediately below the centre of the left eyepiece. Look vertically down into the stereoscope with both eyes open and introduce a pencil between the two right-hand legs of the instrument. Move the pencil towards the left until its point appears to coincide with the dot. Mark this point, and remove the stereoscope. Two dots, some distance apart, should be present on the paper. Measure this distance and repeat once or twice. The distance so found represents the inter-ocular separation of the viewer and is the distance apart that identical objects on stereo prints have to be placed, in order to achieve proper stereoscopic effect.[1]

Stereo examination of prints. (1) Select a pair of stereo prints on which some well-defined detail is common to both.

(2) By the method earlier described, mark the principal point on each and transfer each point to the other print.

---

[1] The inter-ocular distance for most persons lies between $1\frac{3}{4}$ and $2\frac{1}{2}$ in.

(3) Join the principal points and obtain the line of flight.

(4) Lay one of the prints on the table so that shadows fall towards the viewer. This should always be done under good illumination to the rule: 'reproduce the lighting conditions of nature'; for example, by using a shaded lamp illuminating the print as the sun did the object at the time of photography. Within limits shadows may fall towards the viewer's right or left shoulder, but should never fall away from him. Wrong orientation of shadow causes 'reversed stereo', i.e. hollows look like hills and hills like valleys. (This reversed stereo may also be developed if the stereoscope is placed at right angles to the line of flight instead of parallel to it, or if prints are viewed in wrong sequence.)

(5) Superimpose the second print onto the first one so that common details coincide in proper sequence.

(6) Pull the second print away from the first along the line of flight until the two images of the same object are spaced at the interocular distance.

(7) Place the stereoscope over the prints so that the eyepieces lie along or parallel to the line of flight with one eyepiece over each print. Look vertically downward through the eyepieces when the two images will be seen to fuse into one image with full stereoscopic effect. Fix the prints so that they do not alter their positions.

Stereo perception may be increased by *increasing* the eye base in order that some topographic feature may be emphasized. For this purpose the examination of consecutive prints 1, 2, 3, etc., is changed so that 1 and 3 or 1 and 4 are used as pairs; provided, of course, that the specific detail is common to both prints.

There is obviously a lot more in stereoscopic study, but if the reader is content to accept the results of his observations without worrying too much about optical or other reasons, the above simple drill will be adequate for most air photo reading for soil purposes. After the reader has done the drill a few times he will learn by experience to separate his prints to inter-ocular distance, and to match his prints by direct selection of detail. In practice the inter-ocular distance changes somewhat with fatigue, and in some cases reverse stereo perception may be obtained. It is a sensible precaution therefore to limit one's time with a stereoscope. It seems that most interpreters like to change their eyes to something else for ten minutes or so after about twenty minutes of intensive stereo reading.

A PHOTOGEOLOGICAL MAP, MUSCAT COAST

— — — — Bedding in exposed rock

—.— —.— Marker horizon separating marls with thin limestone in the upper part, and thin bedded marly limestones formations in the lower part of the photograph

⊤ Strike and dip of beds

⊤ Dip 0–5°

⊤⊤ Dip 5–25°

-⊖- Gravelly outwash fan deposits of lateral wadis

∿ Sandy and gravelly main wadi bed deposits

ɰ Talus slopes

6 Beach deposits

▨ Salt plug

v Village

PLATE 3

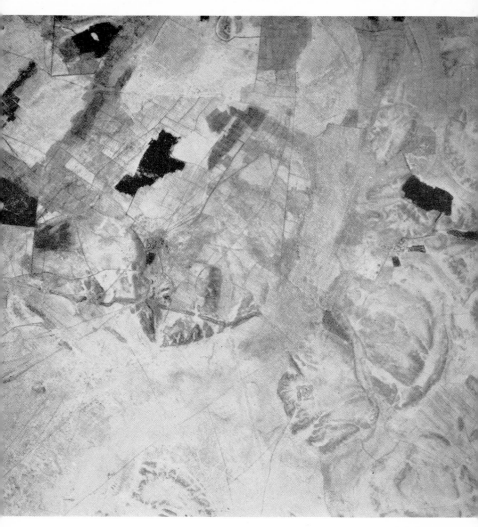

A soil map of the margin of gravel mounds (Lower Fars Beds) where the alluvium of the Adhaim Fan ceases. Ishaqi Project, 30 miles NW of Baghdad, Iraq

Soils: Beled Soils—saline-solonchak soils
Saline facies; Electro-conductivity greater than 15 mmhos/cm   B
Moderately saline facies; Electro-conductivity from 8–15 mmhos/cm in top or middle 50 cm   (B)

Masud Soils—solonetzic soils associated with uneven gullied surface. Strong columnar or prismatic structure   M
Moderately leached facies: Electro-conductivity less than 8 mmhos/cm in the top 100 cm. Saltier substratum. Weakly columnar or prismatic structure   (M)

Wadi Tharthar Soils—variously saline. Undifferentiated sandy soil between gravel mounds. Gypsum content often high   Th

Gravel Mounds—becoming predominant to the west where they form the limiting boundary

Key· ⊙ 652 1½ metre borehole
△ 412 1½ metre Pit
⊕ 671 5 metre Borehole

PLATE 4

Stereo vision without stereoscope. This accomplishment is very useful in the field when it may be difficult to lay out prints and use the stereoscope. One of a stereo pair of prints is held in each hand and the eyes are focused on to some distant object. The eyes are kept steady in this focus while the prints are slowly interposed to cut off the view. The eyes are not blinked or changed so that the viewer appears to be looking *through* the prints. He will then see that his detail has fused and he has got stereo perception. It takes a little practice to become adept at this. Unlike the use of the stereoscope which is supposed to be beneficial to the eyes, this method is apt to make the eyes very tired. It is a useful field trick, none the less.

AIR PHOTO READING AND AIR PHOTO INTERPRETATION

There are two distinct phases in the study of air photos. Air photo reading may be defined as 'the simple identification or description of photo images without analysis of their meaning', while photo interpretation determines the significance of photo images for some special purpose. For example, a trained photo reader becomes an air photo interpreter when he makes a soil map from air photos by using his knowledge as a soil surveyor for the interpretation of the external features of soil he observes on the photo.

The first thing which strikes the attention of the photo reader is a perception of *tone*. Tone is the basis of the photo image since, by its different shades of greys, it gives the outlines of size, shape, shadow, pattern, and position of all the external or associated soil features on the print. All shades of grey appear in the make-up, depending upon how much light is reflected from the object to the camera. Since cameras admit more light through the centre than through the margins, tones around margins are often darker than in the centre. It would seem that the interpretation of tone can only be acquired by actual experience. Many efforts have been made to make grey-scale keys, but they have not been satisfactory because varying factors can make the same object appear in different tones on different prints, or make different objects appear in the same tone on the same print. Since the absorption or reflection of light depends upon the nature of the surface of the object, the photo reader may make a shrewd analysis of what he sees, for example wet patches in a field are of a different tone from the dry patches. Tone in general may be looked upon as the major result of the influence of air photo 'texture'.

Texture in air photos refers to micro-characteristics too small to be identified individually but which when taken together are responsible for a visible result. For example, an arable field in ripe corn appears in light tone. This is made up in fact of millions of small reflections from individual plants; and if a wind is blowing, mottles or stipples darker or lighter in the texture will be seen. An important characteristic related to tone and texture is shadow.

Shadow. Shadow always appears in dark tones, and is the chief factor for the diagnosis of shape and size. It is of particular value for the identification of height and species in vegetation, while illumination and shadow define relief. The best times for taking photos for shadow effects are mid-morning or mid-afternoon when light is quite good, and the sun at such an angle that shadows are long enough to assist in identification and yet not so long as to obscure other detail.

Shape and size. The shape of an object on the print is determined by what is known as its tonal edge gradient (which is the degree of abruptness of the tonal change between the edge of the image and its background of adjacent images), by the angle of view, and by any distortion due to photography. The estimation of the true height of an object is a matter of elaborate technique with photogrammetric instruments and calculations, but in ordinary soil work heights may be sufficiently accurately assessed by direct comparison with a known object or by intelligent interpolation. The length of a shadow is governed by the topography over which it falls. A tall tree with shadow uphill will look like a short tree, while a short tree with shadow downhill will look like a tall one.

One may next turn to the measurement of objects on the ground, for example, width of streams, shape and size of fields and hedges, breadth of tree crown, etc., all of which build up a framework for the diagnosis of the soil properties. There is a limit to which the eye, even with magnification, can detect objects, but it is generally possible to recognize certain details as follows:

At a scale of 1/20 000 an object of 6 ft in length may be perceived, e.g. cattle

At a scale of 1/12 000 an object of 4 ft in length may be perceived, e.g. sheep

THE USE OF AIR PHOTOGRAPHS IN SOIL SURVEY

At a scale of 1/8 000 an object of 2ft in length may be perceived, e.g. poultry

At a scale of 1/6 000 an object of 1 ft in length may be perceived, e.g. furrows

The magnification of the pocket stereoscope ($\times 2\frac{1}{2}$) is a handy scale for general purposes. Magnification beyond certain limits, usually considered to be about $\times 4 - \times 5$, is rarely helpful since the grain of the print gives a woolly indefinite effect. It is for this reason that a large-scale photo is of greater value to the soil surveyor dealing with small and complicated areas than is a magnified small-scale photo.

Photo-measuring magnifiers usually consist of a short cylinder. At the top is an adjustable magnifying eye-piece which focuses upon a graduated scale inscribed on a transparent base plate. Various models differ in the degree of magnification but it usually exceeds that of the pocket stereoscope. Such instruments are particularly useful for the concentrated study of small images in a small field of view. In use the magnifier is placed over the image of the object to be measured, focused for the scale, and the measurement made. The scale of the print must be known, when the length of the object can be calculated by

$$\frac{image\ length\ in\ mm\ \times\ R.F.\ of\ print}{304 \cdot 8} = \text{real length in ft.}$$

Example. Image 4 mm, scale 1 : 10 000. Real length = 40 000 mm = 131·2 ft.

If a magnifier is not available any simple instrument capable of measuring length may be used, but without magnification it is difficult to subdivide one millimetre by eye.

The chief difficulty in measuring length, however, lies in the fact that the scale given on the titling strip does not necessarily give the real scale at any particular point on the print. The scale therefore must be found by some other means, either from data obtained on the ground or from an accurate map. Scale may be determined by measuring the distance between two well-recognized objects on the print. Any units of measurement may be used so long as they are the same for both readings. The R.F. of that portion of the print is then given by

$$\frac{\text{distance on print}}{\text{distance on map}} \times \text{R.F. of map.}$$

In taking measurements from the map, care should be taken never to use conventional signs (e.g. church ⚲), nor objects printed in different colours. The scale may also be found from a ground check, bearing in mind that distance must be measured in plan and, as before, both print and ground distances must be expressed in the same units:

$$\text{R.F. of print} = \frac{\text{distance on print}}{\text{distance on ground}}.$$

Tilt affects the variation of scale over an air photograph. It is produced if the camera is not perpendicular to the horizontal plane of the earth. It is produced in two ways, first by intention, as in the case of 'camera fans' where a small angle of deviation may result in 'near verticals' or the angle of tilt may be so large as to give 'high or low obliques'. Secondly, the tilt may be accidental such as could be caused by the unsteadiness of the aircraft, etc. When the aircraft dips either wing the camera will tilt laterally to give lateral or 'x' tilt, and when the nose of the aircraft tips up or down the camera will tilt forward or backward to give longitudinal tilt, 'tip', or 'y' tilt. Tilt causes a circular object on the ground to appear as an ellipse on the photo, while a square object would appear as a trapezium. Accuracy of scale cannot be guaranteed when distortion of the image, due to tilt, is apparent. Negatives distorted by accidents, etc., are usually useless and are destroyed. The scale of a tilted photo changes regularly over the whole photo, and if the scale at the centre is correct then the scale on the *uptilted* side will be *smaller* while that on the *downtilted* side will be *larger*, i.e. tilt compresses the images on the upper side of the photograph and expands them on the lower side. Scale does not change along any line that is perpendicular to the direction of tilt. For ordinary soil survey tilt less than 2° may be ignored.

Pattern. Spatial combinations of otherwise unrelated features (both natural and artificial) in a landscape are probably the first thing perceived on an air photo and are referred to as 'patterns'. It is possible to learn a great deal about a soil from a study of its utilization pattern alone. For example, there is the paddy field pattern, the water meadow pattern, and the pattern of 'stone wall' country, etc., while artificial drainage patterns can be picked out with ease. It must be remembered, however, that such patterns are to some extent the result of man's interference with the normal course of nature. In fact, clearly defined boundaries are not natural. The limits of the pattern due to a catenary sequence of soils, for example, may be quite clearly defined on an air photo, but the boundaries of the actual soil associates or series in the sequence may frequently have to be determined by some arbitrary division on the ground.

Resolution of detail and resolving power. The quality of the detail on a print is controlled by many variables such as the lens, film combination, movement of image, haze, brightness, contrast ratio, and on scale, i.e., on the size of the image which is being recorded. Resolving power is defined in the *Manual of photogrammetry* (American Society of Photogrammetry 1952) as: 'a mathematical expression of lens definition, usually stated as the maximum number of lines per millimetre that can be resolved (that is, seen as separate lines in the image)'. One unavoidable difficulty results from the movement of the camera during flight because, no matter how quickly the shutter works, the aircraft travels a certain distance during the exposure time and so some degree of blurring becomes inevitable. Sanders (1941) has shown that the range of tolerance of image movement for specific forms of military interpretation may be classified in three standards:

(1) For the best interpretation, image movement must not exceed 0·0135 in.
(2) There is a tolerance range from 0·0135–0·028 in within which the print is declared to be 'pictorially fair' and is still quite good for most interpretations.
(3) The range between 0·028 and 0·042 in renders pictures progressively poorer in quality, though they may still allow fairly accurate application.

For soil survey in general almost any image movement up to 0·035 in is tolerable, though it is obvious that the better the photo the easier it is to interpret. For special purposes such as the classification of woodland, Spurr (1948) considers that image movement beyond 0·002 in is not tolerable.

For the estimation of image movement Sanders (1941) gives the following formula:

$$M = \frac{1·467VT}{S},$$

where M = image movement;
V = ground speed of aircraft in miles per hour;
T = camera speed in seconds;
S = scale of photo in feet per inch.

The scale and the size of the object will of course also affect the resolution, some objects being too small to be recorded on the air photo. Since the human eye cannot perceive objects of diameter smaller than about 0·001 in, while the resolution of an ordinary air photo lies between 9 and 15 lines per mm, it follows that the resolution of the photograph is much greater than that of the human eye. A magnification of $\times 2\frac{1}{2}$–$\times 4$, however, will bring into human perception the smallest detail actually recorded on the photo. No amount of magnification can produce a detail which does not exist on the photo; it can only show what is present.

SOME SYSTEMS OF AIR PHOTO INTERPRETATION USEFUL FOR SOIL SURVEY

There are two main systems whereby the photo reader or interpreter may delineate soil units on the map. He may use one of the numerous 'key' systems or he may use the 'convergence of evidence' technique.

Keys. In key methods evidence previously obtained and recorded by scientific specialists skilled in the interpretation of air photos is recorded in a form intelligible to other persons who are not necessarily specialists. For example, a geologist making a key would know from personal experience what an image on a photo really represents on the ground and its diagnostic value for the particular task in hand. Such information is collected in the form of annotated prints, essays, or legends and incorporated into an indexed key for the use of less highly-trained photo readers. A photo reader using a key may never have seen the real object at all, but recognizes its image because he has been trained to do so.

The simplest form of key is dichotomous, based on a process of elimination so that the key-user follows a prescribed step-by-step process whereby one of two alternative deductions is eliminated at each step, and the other is bifurcated until final identification is achieved. Keys of this type must be very skilfully designed, and must be accompanied by an exceptionally well-written text, so that each step is as nearly foolproof as possible, because a faulty deduction at any step will lead further astray at each succeeding step until the final deduction is completely absurd.

To offset this difficulty an *associative key* sets out in essay form characteristics relevant to the interpretation but not normally capable of diagnosis from the dichotomous key. Another type of key known as the *selective key* may be very useful for solving a particular problem at any stage in the air photo reading. Here the photo reader selects the example from a library of annotated prints that most closely corresponds to the image and environment he is attempting to identify. This trial and error method may, however, cost much time.

There is yet another kind of 'key', somewhat different in character from those previously described in that it is made by one specialist for the use of other specialists studying the same subject. It is particularly adaptable for certain aspects of soil survey such as the classification of catenary sequences and their complements of associated soils. Such a key is the *analogous area key* and consists of a collection of annotated prints with a corresponding legend describing in technical terms all the environmental features and soil characteristics of a surveyed area that the soil surveyor would find of benefit in planning the survey of a new but similar area. From the study of the key of one soil region the surveyor can learn to recognize analogous detail on air photos of a new region and may tentatively draw soil boundaries directly on to the map without the intensive ground survey that had to be done for the mapping of the first region from which the key was devised. Boundaries of this nature can only be drawn from the interpretation of images perceivable on the print, so that the use of the key still does not preclude the need for confirmatory ground survey, or for the interpolation of boundaries that lack external expression and are detectable only from borings. This technique is particularly useful to the soil surveyor using air photos to speed his ground work.

For the more experienced photo interpreter another technique may be used. This is the *convergence of evidence method*. Instead of the 'break-down' techniques previously discussed, this depends upon the building up of many different observations and deductions to converge on to but one conclusion. The advantage of this method is that one faulty or possibly anomalous observation need not necessarily invalidate the whole deduction. For instance, if six observations lead to a certain conclusion while one observation does not seem to 'fit' there would be a case for ground check to investigate as to whether or not this anomaly possessed any particular significance of its own. Such a contingency might arise, for example, where all the diagnostic vegetation of a furze heath had recently been destroyed by fire and was just beginning to regenerate.

The use of air photographs in the field

Air photos have great value during 'indoor preparation for mapping'. The photo may also be used in the field, either for accurate location finding or as a base map for direct mapping. The use of stereo pairs is always preferable for field study although it adds a little to the quantity of equipment for the field-man to carry.

The use of prints as field slips is becoming increasingly popular because an up-to-date photo will not only show things that a map cannot reproduce in detail, such as intra-contour relief, position of lone trees, etc. (which are excellent for location-finding), but also changes in utilization boundaries, erosion channels, etc., which have come into being since the map was made.[1] In very flat or open country with very few recorded landmarks, or in difficult country with too many, the photo study is probably the quickest possible way in which to recognize locality, always remembering, however, that the photo has a variable scale. In undeveloped countries for which adequate maps do not exist, the air photo may be the only basis on which the soil surveyor can work in the field. R. H. Gunn (1955) provides an interesting example of the use of air photos during the soil survey of 15 000 miles^2 in the Sudan. His general method of working appears to be that of the *analogous area technique* and is briefly described in the following notes.

(1) The survey was carried out during the two dry seasons between 1951 and 1953.

(2) During the first season a ground survey was made of representative areas, soil samples collected and analysed, and the major groups of soils classified.

(3) Aerial photos were made of some 4000 of the 15 000 miles^2 at the start of the second season, and prints were delivered immediately to the staff in the field. Much use was made of 'stick-down' uncontrolled mosaics.

---

[1] The authors have seen some air photos of a large river that had changed its course by some 3 or 4 miles to cause a complete change in the distribution of the vegetation and human settlement since a map of the area was last made about twenty years ago.

(4) The scale of photography chosen was 1/25 000, this being considered the smallest possible for the interpretation of the soils and correlated vegetation within the survey area.

(5) The *interpretation* of the soils was based on a *detailed examination of representative areas* with the *photos in the field*.

(6) It was found that:

(a) Marked soil changes, chiefly in texture, give rise to changes in vegetation, for example acacia woodland occurred on the lighter textured aeolian and alluvial soils, while scrub or sparse annual grasses occurred on the clay soils.

(b) Changes in vegetation also marked areas which were seasonally flooded by rivers or by rain.

(7) In *cultivated* areas it was found that the better-drained higher-lying ground was being used for sorghum and sesamum while the lower-lying depressions, often saline, remained uncultivated.

(8) Soil boundaries could be mapped from a knowledge of the relationships between soils, parent materials, and land forms.

(9) *Tone* was found to be of great value for boundary definition, for example sandy soils were indicated by a white or very light tone while clays and seasonally flooded soils were indicated by dark tones. Local ground-check was needed in certain clay areas covered by dry grass, which produced a similar tone to that of sand and could be mistaken for it when reading the air photo.

(10) Finally, after the completion of this field work, the major part of the air photo interpretation was carried out in the office, matching each area to one of those already examined.

The soil and geology map overlays, legends, and air photos of Plates 3 and 4 (pp. 108–9) illustrate how air photos may be used.

Soil maps for military purposes

For military purposes there are three distinct requirements: first, the ability to make a soil map from a combination of ground study and air photo interpretation in accessible areas; secondly, the ability to make a soil map of inaccessible areas; and, finally, to be able to translate the information obtained into a legend capable of being used by any branch of the Services. The making of the legend is probably the most difficult task the photo interpreter has to do.[1]

[1] See Fig. 13, pp. 118–19, an experiment in the making of a soil-map legend for military purposes from the study of air photos of a site assumed to be inaccessible.

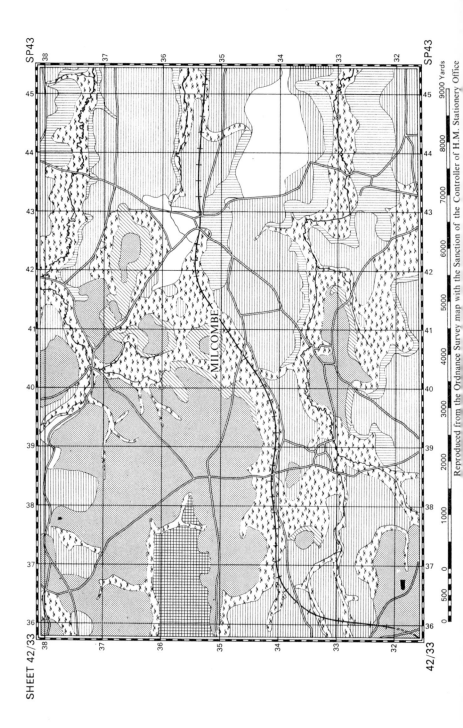

MILCOMBE

9000 Yards

8000

7000

6000

5000

4000

3000

2000

1000

500

0

Legend to the soil map

Pedological classification	Map symbol	Military equivalent	Derivation of soil	Military characteristics for 'Going' or Airfield Construction
Podzolic Soils (Leached Sands) (Tadmarton series)		Sands with excessive percolation	High level, leached Northampton Sand in situ	Good going at all times. Requires stabilizing for aircraft. Usually heath or reclaimed heath to poor arable.
(1) Brown Earths (Low base status) Sands		Sandy slope soils well drained, liable to slip downhill	Colluvium of sand washed down over marlstone or clay	Good going at most times but spring lines occur at junction of sand with clay. These flushes may cause bogging. Topographically unsuitable for airfields.
(2) Brown Earths (High base status) Loams		Loamy slope soils. Good surface drainage; imperfect at depth	Colluvium of marlstone loam over clay; Lower Lias	Good going at most times but liable to wetness at junction of loam on clay. Good arable or excellent perm. grass. Topographically unsuitable for airfields.
(3) Brown Earths (High base status) (Loam of Banbury series)		Plateau loam on firm rock at depth; well drained	Middle Lias Marlstone weathered in situ	Good going at all times. Excellent arable, rarely in perm. grass. Suitable for airfields
(4) Rendzinas (Red and Brown Calc. Soils) Hogs Norton series		Shallow Brash loams on limestone rock; well drained	Gt. Oolite Limestone weathered in situ	Good going at all times. Good general arable practice. Suitable for airfields.
(5) Gleys Barford series		Clays and heavy soils with poor drainage	Colluvium and alluvium of riverain and valley-bottom sites. Lower Lias Clay	Poor going during wet seasons. Mainly meadow land, frequently liable to flood. Only suitable for advanced temporary airfields in dry weather
Undifferentiated				

In the author's opinion the 'convergence of evidence' technique specifically adapted to the recognition of the characteristics of 'analogous areas' is the best method. In the first place, therefore, the trained interpreter must get to know something about soil by actually walking upon it, with his prints in his hands, and learning all he can about the associated features of the particular site and soil on which he is standing. There is no other way than this of discovering the significance of micro-relief, texture, ground vegetation, etc. For the recognition of *soils in inaccessible areas* the interpreter must have been through this first training stage and must have built up a 'mental library' of analogous areas, so that he can say: 'I've seen that before, that is so and so.'

The interpreter will use also whatever knowledge he may be able to obtain from any source at all, such as geological maps, tourists' guides, etc. It is well to remember, however, that geological maps are not soil maps: they should always be used with the greatest discretion. *Solid* geological maps show the underlying massive strata and though these frequently tie in with the land forms they do not always show the material from which the overlying soil is derived. *Drift* geological maps show the surface deposits to some arbitrary depth (very shallow drifts are ignored) and, though these deposits are frequently the parent materials of the overlying soils, drift maps are not, nor were ever intended to be, soil maps. Another important point to bear in mind is that the parent material is only one of the five factors that operate in soil formation; it is the functioning of the other four factors as well that determines the kind of soil that will develop from exactly the same kind of rock under different influences.

For example, an igneous rock weathering to form a soil in the tundra region under the influence of Arctic conditions of temperature, humidity, etc., and very restricted chemical and biological processes, will produce a shallow frost-shattered and stony skeletal soil with impeded drainage in equilibrium with mosses and lichens. Under tropical conditions of high temperature and seasonal rainfall, with intense chemical and biological reactions and free drainage the identical rock weathers to form another kind of soil, a deep red tropical loam. Or if the soil site is of the 'water-receiving' type (i.e. basin or plain) with impeded drainage then a black tropical soil in equilibrium with grasses may result. This same igneous rock will produce different soils in all parts of the world depending upon the varying effects of the other four soil-forming factors, though the geological map would still show the rock as being igneous. Thus, to recognize soils of inaccessible areas by the study of air photos the interpreter must first know the conditions under which the materials shown on the map have broken down and developed into soils. From some source or other he must obtain information about the climate, relief, physiography, and the mass and character of the vegetation.

In the early stages of building up his mental library an identification scheme such as that given in the *Soils guide* (Table 9, p. 122–5) may be used. Though by no means complete this gives some idea of how the problem of identifying soils in inaccessible areas may be attacked. It assumes that the interpreter is using air photos at 1/10 000 to produce a soil map at 1/25 000.

The following points illustrate how some of the categories given in the *Soils guide* may further be developed to stimulate the reader to evolve his own mental library.

Bare solid rocks. These may show as massive or layered structures. Acid rocks, i.e. those with a high silica content such as granite, flints, or sandstones, give light or white tones, while basic rocks such as basalt give dark or black tones.

The shapes of the exposed rocks often give useful clues: for example, owing to the even distribution of their particles, sandstones, chalks, and clays tend to weather evenly to form bold rounded hills. Well-bedded sedimentary rocks tend to form ridges with a steeper slope on one side than the other and with well-defined denudation patterns and river systems.

Shallow soils on rock. Shallow chalk soils on solid chalk may be recognized by

(1) Bold rounded features with dry valleys;

(2) Few trees, though juniper and yew are often very clearly defined. Short turf generally, but irregular patches of white grass may sometimes occur, particularly if the land has recently fallen to grass after arable use.

(3) Fields (if any) large and devoid of running water, hence there is little need for hedgerows or ditches. On slopes less than 15° ploughed fields show pale grey in the field and give a very light tone on the print. As slopes become more gentle towards the bottom of the feature, dark browns appear in the field with a gradual darkening

The identification of soil categories by th

SOIL CATE-GORY	INDIVIDUAL FIELDS		VEGETATION				MICH
	SIZE AND SHAPE	BOUND-ARIES	TYPE	HABIT AND QUALITY	MICRO-RELIEF	PITS AND QUAR-RIES	RIVEF BANK
			Attempt to estimate height and canopy				
Texture cat. (1) Heavy (Clay and silts)	Clay hills or variable topog. Small and ugly, irregular. Ugly fields also occur on sites on sands or rocks but use other criteria. Clay plains give large fields with good hedges.	Ditches and hedges: artificial straight, natural crooked. Follow the drainage pattern for run-off or spring lines.	Hawthorn, ash, hazel hedges. Oak and ash as hedgerow timber. Permanent grass is very common. Woodlands are usually deciduous.	Isolated trees in fields. Small clumps may conceal ponds. Hedgerow timber frequent and of good quality.	Generally flat but on slopes surface run-off shows clearly. Ridge and furrow well defined. The closer the ridges the greater the need for surface drainage. Ridges usually some multiple or fraction of a chain of 22 yd. Water meadows. See cat. (5)	Well-drained clays may show dry pits but this is very rare. Pits are nearly always wet. Ponds are common and often hidden by trees, e.g. willow. Brick clays often show kilns and drying yards.	High inf good drainage Moderat moderat drainage Low infe poor drainage often wit permane high wat table. Check general view for flood pla pheno-mena.
Texture cat. (2) Medium (Loams)	Generally large and regular but ill-drained loams often show up by artif. drain. system. Dominantly arable.	Good drainage: hedges banks. Poor drainage: ditches hedges	Hawthorn, hazel with sycamore, oak, and elm as main standards. Compared with clays far fewer hedgerow trees. Shade not desirable over crops.	Isolated trees in field rare. Hedgerow timber usually good. Large woodlands not common as soil usually too good for trees. Small estate coppices and amenity woods may be found.	Rarely dead flat. Usually positive, gentle slopes. Ridge and furrow rare or absent. 'Lands' show up as drainage furrows usually more than 22 yd apart.	Marl pits, Chalk pits, Limestone, &c,. can usually be detected by tone, usually dry. Gravelly loams often around wet pits with washing machinery.	As for cat. (1), b check fo terraces above th flood pla These wi usually b well-drained topo-graphica units.

erpretation of air photos without access to site

ATURES					
OADS AND RACKS	GATES AND STILES	BARE SOIL OBS.	SPEED OF PERCOL. DRAINAGE	GENERALIZATIONS Examine for evidence of:
ding often many ges of of cles g drier nd. ld s are ght are lly and lled. p- ed s lly fol- cial nage ern.	Usually well marked by cattle tracks and ruts, expos- ing puddled soil and reflecting much light. Tie in with ditches and particular- ly culverts. Under magnif. the structure of gates and stiles show local wood, i.e. 5-barred gate, &c.	Animal tracks usually well defined to ponds or steading. Recently ploughed land shines silver and mottles of raw clay show where subsoil is disturbed. Under- ground tile or mole drains show up for a very long time by difference in tone.	Virtually nil on slopes, slow on flats. Look for run-off gulleys. Impeded or imperfect, with pro- file gleyed.	Standing water. (This is bad because *sub aqueous* observations are impossible.) Surface-water gleyed soils or temporary flooding seasonally wet. Selective grazing. (Patches of irregular grazing show wet and dry or good and bad crops. White grass and rushes not eaten.) Identification of animals. (Sheep are rarely long on wet soils. Folded sheep are good indicators as a rule for well-drained soils.) (Store cattle often indicate good strong fairly well-drained soils.) Ricks, bales, cocks, sheaves, etc. Size and number per acre may be measured and give an index of soil quality. Clays on hill-sides are frequently scored by abrasives washed down from coarse material higher up the hill. Buildings, size, shape, and disposal very important. Barns arable. Stockyards grass. Lack of buildings shows perm. wetness, etc. Animal tracks going to gates or ponds show up and betray drainage. Where large numbers of cattle show in one field or so, this may be temporary or milking and is not of much significance, but if a whole block of the country is characterized by cattle then good feeding is indicated, strong loams or clays. Old buildings are frequently built of brick. *New* buildings of brick are not so significant. When a spring line occurs on one side of a hill or feature and not the other it may be assumed that the substrata dip towards the springs, and flushes will show the junction of the strata. Particularly sand over clay in rolling topography.
ally defined fairly ght. nges of e not mon if tracks ====B ows onal ess es slip- going ubber- i les.	Sometimes bare soil but absence of much rutting carrying far into the field. Tie in with local drainage. Gates and stiles of local wood. Modern fences often wire with concrete posts.	Loams are usually very much more uni- form in tone since subsoil is usually deeper and of similar colour. Arable practice dominates the tone of photo. Check by compara- tive season- al cover.	Generally free to im- perfect, i.e. moderately fast to slow, never excessive. Usually percolation exceeds surface run-off.	As above, but also: Attempt to recognize system of cropping. These soils are more often in arable than in permanent grass. The variable nature of tone in different fields on same kind of soil growing many different crops is indicative of good soil of good texture, workability, and drainage. Buildings may be of brick, stone, or wood roofed with thatch, tiles, or slate. Half-timbered buildings on good soils. Medium and light loams are very liable to frost-lift and when thawed often look much wetter than they really are. Some farmers often build ricks of bales that re- semble cottages with an up-ended bale as chimney. These may be detected by the lack of gardens and paths or outhouses. This *cannot be done with hay bales* which will not stand right.

SOIL CATE-GORY	INDIVIDUAL FIELDS		VEGETATION		MICRO-RELIEF	PITS AND QUAR-RIES	RIVI BAN:
	SIZE AND SHAPE	BOUND-ARIES	TYPE	Attempt to estimate height and canopy HABIT AND QUALITY			
Texture cat. (3) Light (Sands and gravels)	Large and regular when under arable but large tracts of land un-fenced are often to be found with patches of heath, bare ground and pits on siliceous soils.	Sparse hedges. Tendency for banks and dry ditches.	Plantations, commons. Vaccinium, gorse, Rubus, and bracken heath. Birch and Scots Pine natural. Broad-leafed trees generally sparse and poor. Riverain calcareous gravels much resemble cat. (2).	Grass rare except on very poor sites. When usable as arable is often in market-garden use. Very dark tone (do not confuse with wet-ness.)	Ridge and furrow absent. Catch drains at spring lines over heavy subsoil. Often gent-ly terraced or flat. High and dry gravels no drains. Low and wet gravels ditched.	Sand and gravel pits both wet and dry show clearly. Siting of pit often decides if it is sand or gravel. Generally river gravels are low and terraced while glacial gravels are on upland sites.	As for cat (1) for cat.
Texture cat. (4) Shallow Soils on rock (Skeletal: various textures)	Large on plateau and plain sites but often ir-regular and small in broken topo-graphy. Soil depth often shown by plough when very shallow.	Hedges sparse. Usually stone walls or banks. Very shal-low soils cannot be drained but deeper and ill-drained ones may have ditches.	Almost any type de-pending upon parent material.	Height, growth, and crown should be studied for info. on nature of rooting,	Bare rock often pro-trudes and micro-ero-sion is common and is easily spotted. Very rocky sites are very ir-regular and often give a mottled photo.	Pits and quarries are common. Look especially at hill-side sites for depth of over-burden. Nature of rock can often be found by way in which quarry is worked and tone of photo.	Riveral sites un commo Look f type drainag natural land fo This als helps t denote of rock
Texture cat. (5) Alluvium River silts Peats and bogs	Fields usually on riverain sites. On wet waste areas size and shape mainly con-trolled by drains. Peats and bogs by topography and drain. Artificially drained peats usually rectangular with hedge-less ditches.	Boundaries tend to follow con-tours and river terraces. Hedges often wide with tall timber, Lombards, willows. Peats and bogs. Hedges and trees sparse.	Riverain sites. Natural alder and willow. Single trees here and there in meadows. Wetter patches sedges, etc. Peats. Plantations of conifer or birch. Exposed trees often deformed. Blanket mosses on wet heath.		Flat or terraced on riverain sites. Water meadows can be spotted by layout and service gates in ditches. Peats often in broken micro-relief with patches of water and raised peat as little mounds.	Riverain sites give wet gravel workings with small overburden. Peat cut-tings and stacking show very clearly.	Riveral sites; lc banks s high wi table. Peats a bogs m show st torrent banks.

ATURES				
OADS AND RACKS	GATES AND STILES	BARE SOIL OBS.	SPEED OF PERCOL. DRAIN-AGE	GENERALIZATIONS Examine for evidence of:
ally de-l and ight on ned . Often ken ls with banks. en, how-ls are with and ace on ls. ally geless in culti-d areas, icularly sea.	Often marked by bare patches. Rutting is not common and tracks do not last long. Due to shortage of wood, gates and stiles are poor or absent. Wire and concrete frequent.	Animal tracks, poultry scratches, rabbit warrens all show clearly. Gravels often show as large, bare patches, while sands are dotted with much more vegetation. Flints and glacial gravels show a bright white tone.	Usually little run-off. Percol. is rapid to excessive. Siliceous sands and gravels tend to podzo-lize to give imperme-able pans. See cat. (5), peats.	As above, but also: Look for evidence of wind-borne sand accumulat-ing to dunes or covering roads and tracks. Look for sheet or gulley erosion. Soils most likely to erode are coarse textured with a very limited particle-size dis-tribution and with little or no vegetational anchors. Erosion marks may often be seen on a clay slope, but investigation will show that the clay has been abraded by some coarse-grained material rubbing over its surface. Buildings are often poor, of brick with slate or tiled roofs. Wood and stone are rare. These very permeable soils usually slow a poorly defined drainage system, unless artificially drained, when catch drains may show on slopes over heavier soils. (Catch drain—interceptor drain.)
cks n con-ed by graphy. rocks n give ken ls.	Often good with stone-pillared gates of iron. Stiles of stone not wood. Check nature of bridges and culverts.	Hard rocks often show boulder-like lumps. Soft rocks, chalk, lime-stone, often show changes of tone on photo especially on arable or where animals disturb scanty herbage.	Surface soil usually fast but may often be held up by imper-meable rock beneath. Look for seepage.	As above, but also: These soils are usually so variable, due to the near-ness of the parent rock, that mottles are common. Use every point you can think of and remember that wet patches or dry patches can show up with varying depth of soil. The whole general 'lie of the land' must be studied. Buildings: Chalk gives brick, wood, or half-tim-bers with roofs of thatch, tile, or slate. Limestone gives stone or wood or half-timbers with roofs of thatch, tile, or slate. Hard rock gives stone or brick or half-timbers with roofs of tile or slate. Bare chalk, weathered flints, or certain sands and gravels some-times look alike in tone but, when other edaphic factors are considered, there should be no danger of confusion.
en wind con-cks n tic by osing er and. alled ls ally very ight sharp ds dered litches. verts are uent on roads. ds er than	Very un-common.	Nearly always shows bare patches of soil in bog country, boulders, etc., stand out. Riverain sites show black in gateways, etc. Bare soil is rare except where artificially dug drains may show white, e.g. podzol.	Organic surface has high water-holding capacity and though percolable may still be wet. High water-table is general and percolation controlled thereby.	As above, but: Many features resemble clays of impeded drainage but other features are usually so outstanding that confusion need not occur. There is usually a pronounced lack of buildings and topography is nearly always the most obvious guide.

of tone on the print. Chalk grass is light in tone, with every scar or track standing out white on the print.

Shallow stonebrash soils on limestone rocks. (1) Ridges tend to be more steep on one side than the other with streams in valley-bottoms. Streams frequently have a cliff-like bank of solid rock on one side with meander belts and alluvial flats on the other.

(2) Trees are generally more frequent than on chalk, with gorse patches frequent on overdrained sites. Slopes are often grass-covered with definite sheep-walks or lynchets or, where soils are deeper, well-wooded slopes of broad-leaved trees.

(3) Where limestones are fairly soft and the soils are not too shallow, field boundaries are usually hedges. Where rocks are harder and the soils are very shallow, stone walls are dominant. Fields are usually large and regular on the broader uplands but in broken topography they may be small and irregular. Ditches rarely occur except in lowland sites. The tone of the soil in the field is red or brown to give darker greys on the photo than is the case with chalk but the grading of lighter to darker tones as the soil deepens downhill is similar to that of chalk. Although stonebrash soils are considered to be 'dry' soils, they are not as 'dry' as chalk.

Clays. (1) Clay hills are generally smooth and round; unless the surface has been abraded by surface wash carrying coarser (abrasive) material, they are peculiarly free from erosion gulleys. Clay plains are usually characterized by rivers with low but clearly defined banks, with meander belts and ox-bows. Usually spring lines do not occur actually *in* clays but are indicative of some more porous material lying *over* the clay. The spring line is a mappable junction between the two kinds of soil.

(2) Woodlands of damp clay areas are generally characterized by oak with or without hazel, etc., coppice. The standards are usually far enough apart for the crowns not to touch (normally about twelve or fewer per acre). The shade cast by such standards is very much less dense than that cast by close canopy.

(3) Fields are in permanent grass and if grazed hard are of uniform tone but if undergrazed may appear as light tone with darker mottles. The banding of ridge and furrow cultivation may be clearly observed. Lone, old trees are frequently to be seen on grass that has never been ploughed and also in the ridge and furrow of old-time cultivation. Careful study of the size, shape, and habit of lone trees can be a very important aid to the diagnosis of drainage

and exposure. The woodlands of the more calcareous clays are characterized by ash, which gives a different tone from oak due to a lightish 'flash' in the leaves. In areas of non-calcareous clay ash is only dominant where the soil is definitely wet, i.e. ash likes a high lime content or a high water content in the soil.

One could continue on these lines for many pages but the point to be made is that the air photo interpreter of soils in inaccessible areas can rarely diagnose the soil from its surface features alone but must use every artifice and preknowledge he can obtain.

6. Some Systems of Soil Evaluation

THESE methods are usually concerned with attempts to map 'land values' or 'fertility' by plotting on a base map a series of isolines for data obtained by the chemical and physical analysis of representative soil samples. The various methods employed to obtain the necessary data differ slightly in different countries, but it appears that for the mapping of soils in the intensively cultivated areas of central and western Europe, the principle is, in general, a sound one.

In Russia use has been made of iso-humus lines and iso-carbonate lines in the mapping of some of the productive regions of the Ukraine, but this is now discontinued in favour of the 'landscape' method.

In England the nearest approach to such methods has been made by advisory officers, etc., in the preparation of plant nutrient deficiency maps during their routine investigations in their provinces.

Certain countries influenced by German soil science make use of what is termed the *Bodenbonitierung* or *Bonität Scala*.

THE *Bodenbonitierung* OR *Bonität Scala*

Numerous systems have been evolved for the evaluation of land, and a good deal of detailed discussion on the subject may be found in Blanck's *Handbuch der Bodenlehre* (1931). The general principle underlying most of the schemes, however, is the evaluation of the land unit by summation of points awarded for a number of factors contributing to the agricultural production of the soil. It will suffice here to take as examples the schemes evolved by Fackler and quoted by Blanck, and the Hungarian system of Kreybig.

Fackler assesses the soil region on three main contributory factors:

> soil conditions,
> the climatic-vegetation complex,
> the economics of transport.

The region is examined and points are awarded on field observation according to the following categories.

The soil conditions

(1) General texture, structure, constitution, and humus.

Best humus loam	30
Sandy loam	20
Sands	9
Chalk	5

(2) Quality of the ploughed layer (structure and humus).

Very good Crumb	10
Good Crumb	5

(3) Quality of subsoil in relation to the topsoil (structure, constitution, depth, etc.).

Very good	15
Good	10
Tolerable	6
Bad	3

(4) Soil moisture conditions.

Very good (perfect drainage, adequate moisture retention, etc.)	10
Good	8
Satisfactory	6
Generally satisfactory but liable to be wet or dry . .	4
Tolerably satisfactory but liable to be too wet or too dry .	2
Continually wet or continually dry	0

(5) Topography.

Level	10
Undulating	5
Hilly or broken	0

(6) Cultivation, manuring, and reserves of plant nutrients.

Very good	15
Good	10
Tolerable	5

Possible marks for soil conditions 90.

Climatic-vegetation complex

(1) Situation of region with reference to optimum conditions.

Wine region	10
Wheat region	6
Rye region	4
Mountain region	0

(2) Danger of hail in region.

Little risk	10
Moderate risk	5
Frequent risk	0

Possible marks 20.

Economics of transport

Distance from railhead.

Less than 5 km	10
Between 5 and 10 km	5
Over 10 km	0

Total possible marks for the perfect region 120.

THE HUNGARIAN SYSTEM OF KREYBIG DE MADAR

This system of soil cartography was employed by the Royal Hungarian Geological Survey in the preparation of their soil maps on the scale of 1:25 000. This abstract was prepared from Dr. Kreybig's notes.

Land value and crop production are correlated with pedological data obtained from site and profile examinations. The 'productive value' of the soil depends upon its fertility, which is estimated from the maximum yields of various plants and the cost of their production. This productivity is the result of the interaction of certain factors that are studied both in the field and in the laboratory, and may be summarized as

(1) climate,

(2) topography with special reference to *micro-relief*,

(3) the geological origin of the soil mass,

(4) human influence, methods of land utilization and treatment by cultivation, manure, etc.,

(5) the soil profile, and the determination of:

(a) the content and nature of the humus,
(b) the depth of the soil used by the plants,
(c) the depth of the subsoil,
(d) the depth, movement, and nature of the ground-water.

Field methods

Inspection pits are prepared on the points of intersection of a network of lines based upon topographical data. These sites are numbered and shown on the base map. The site and profile characteristics are then recorded on the 'field work questionnaire' in the following order:

(1) date and weather,
(2) serial number; depth at which analytical sample is taken,
(3) topography,
(4) stratification [for each layer in order]:

> depth of layer in centimetres,
> moisture conditions,
> colour,
> kind of soil,
> structure,
> pH value,
> calcium carbonate,

(5) depth of humus layer in centimetres,
(6) depth of wheat or corn roots in centimetres,
(7) last crop taken,
(8) special notes: depth of water-table.

The method of assessing the various characteristics and their expression by conventional signs is specific for the country, but it exhibits a strong Russian influence.

Topographical position. Four categories are used:
Depression site,
Slope site,
Elevated site,
Flat site.

Moisture condition. This is determined by the senses of touch and sight:

> 0 air dry,
> + weakly moist,
> ++ moist but will change colour on further wetting,
> +++ no change of colour on further wetting,
> ++++ waterlogged.

Kinds of soil. Under this heading is included the texture, the mode of formation, and to some extent the geological nature of the soil mass. The 'kinds' are diagnosed by feel, and from observations of the reaction to acid and indicators:

	Clay
Meadow	Clay
Loamy	Clay
	Loam
Clayey	Sand
Loamy	Sand
Fine	Sand
Coarse	Sand
	Loess
Sodium	Loess
Magnesium	Loess
Lacustrine	Clay
	Peats
	Grit
	Gravel
	Stony
	Alluvial
	Mud
	Gleyey
	Alkaline

Structure. This item appears to include not only the soil structure as generally understood, but also the general constitution, consistence, and chemical deposits of the horizon:

Crumb (excellent)
Crumb (good)
Prismatic
Polygonal
Laminated
Sandy (probably single-grained)
Dust
Structureless

———

Dense or tenacious
Pitchlike or tacky
Cracks or clefts

With calcareous concretions
With calcareous veins
With calcareous spots
With gypsum efflorescences

With ferruginous veins
With ferruginous concretions
With vivianite

With efflorescences of salts

Depth of humus. This describes the depth to which the intimate humus is distributed in the whole soil mass, and does not include the Hesselman layers.

Calcium carbonate. The presence or absence of carbonate is determined by the violence of the reaction to acid. Four categories are employed:

$$
\begin{array}{lll}
0 & \text{nil} & \\
+ & \text{little} & < 5 \text{ per cent approx.} \\
++ & \text{much} & > 5 \quad ,, \quad ,, \\
+++ & \text{violent} > 10 \quad ,, \quad ,, \\
\end{array}
$$

Laboratory data. Laboratory investigations are extremely comprehensive. The following series of determinations are made, and set out as in the table, which is published as a memoir accompanying the map.

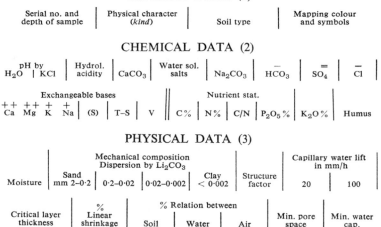

GENERAL DATA (1)

Serial no. and depth of sample	Physical character (*kind*)	Soil type	Mapping colour and symbols

CHEMICAL DATA (2)

pH by H₂O	KCl	Hydrol. acidity	CaCO₃	Water sol. salts	Na₂CO₃	$\overline{}$ HCO₃	$\overline{\overline{}}$ SO₄	$\overline{}$ Cl

Exchangeable bases						Nutrient stat.					
++ Ca	++ Mg	+ K	+ Na	(S)	T–S	V	C%	N%	C/N	P₂O₅%	K₂O%	Humus

PHYSICAL DATA (3)

	Mechanical composition Dispersion by Li₂CO₃					Capillary water lift in mm/h	
Moisture	Sand mm 2–0·2	0·2–0·02	0·02–0·002	Clay < 0·002	Structure factor	20	100

Critical layer thickness	% Linear shrinkage	% Relation between			Min. pore space	Min. water cap.
		Soil	Water	Air		

10

The production of the map

Three conventional characteristics are employed to express detail in the final map:

(a) colouring,
(b) hachuring,
(c) numbering [enumeration points of plant nutrients and ground-water level].

Colouring. Colours are used to show the soil type and class with certain chemical characteristics such as lime deficiency or alkalinity, for example:

Blue. Steppe soils with no lime requirement.
Red. Acid meadow-clay and degraded steppe soils—lime required
Mauve. First class alkali.
Lt. mauve. Second class alkali—and so on.

Hachuring. Superposed on the colours are hachures to denote physical characteristics; for example:

Deep soil with excellent structure and water conditions

Deep soils with good structure and water conditions

Deep, very binding, badly permeable soils

Deep sandy soils, very permeable. Low water capacity

Shallow steppe and meadow-clay soils

Enumeration. A small circle is marked at each site, in which is a series of figures expressed like a fraction. $\frac{513}{80,3}$

The numerator. The first figure indicates the quantity of humus near the surface by the following scale.

1	.	.	.	.	.	.	1 per cent
2	.	.	.	.	.	.	1–2 per cent
3	.	.	.	.	.	.	2–3 per cent
4	.	.	.	.	.	.	3–4 per cent
5	.	.	.	.	.	.	4–5 per cent
6	.	.	.	.	.	.	5–8 per cent
7	.	.	.	.	.	.	8–15 per cent

The second figure indicates the total *phosphoric* acid by the following scale.

1	.	.	.	.	.	0·05–0·1 per cent
2	.	.	.	.	.	0·1 –0·15 per cent
3	.	.	.	.	.	0·15–0·2 per cent
4	.	.	.	.	.	0·2 –0·3 per cent
5	.	.	.	.	.	over 0·3 per cent

The third figure indicates the total *potash* on the same scale.

The denominator. The first number indicates the depth of the humus layer in centimetres.

The second figure [separated by a comma] indicates the depth of the ground-water level in metres.

This method of cartography has much to recommend it, but it must be remembered that the labour and costs entailed in its production are tremendous.

G. R. CLARKE'S METHOD FOR THE EVALUATION OF THE INHERENT FERTILITY OF FIELD SOILS[1]

This method depends upon the ability of a field observer to assess quality classes of soils from the physical attributes of their profiles. In practice the *absolute* value of the soil is not essential; it is more important to obtain some *comparative* values so that one soil series or type may be rated as better or worse for some special form of utilization than another.

It is generally held that the soil attributes that affect plant growth most,[2] and can be measured in the field, are

(a) the texture of the soil,
(b) the depth of the soil,
(c) the quality of soil drainage.

So for evaluation, soil profile pits are dug to a depth of at least 30 in (except where circumstances do not permit, for example on shallow soils on rock or waterlogged soils). The choice of the 30-in profile is to some extent arbitrary but normally few roots of crops are visible below this depth. Texture and drainage are recorded for each horizon, as follows.

[1] G. R. Clarke (1951).
[2] Note that the relative weights to be given to these properties will not be the same in environments differing in rainfall regime.

The estimation of texture value (*T*). Textural types are recognized on 'feel' and inspection of the soil mass at 'field moisture-content'. They are graded as follows.

Field diagnosis	Grading
Gravel	3
Light sand	8
Heavy sand	14
Light loam	16
Medium loam	20
Heavy loam	18
Light silt	15
Medium silt	12
Heavy silt	6
Clay a	15
Clay b	5

It is found in practice that there are two different kinds of subsoil clay: one, called clay (a), is composed of structural aggregates with well-defined porosity, and another, clay (b), is massive and tenacious with little or no visible pore space—i.e. structured and structureless clay material, respectively. The texture generally varies with depth, so it is necessary to obtain an estimate of the texture value over the whole of the profile. This is determined by measuring the depth in inches of each distinct horizon. This measurement is then multiplied by the appropriate texture grading and the whole summed together.

Thus a soil consisting of equal depths of light loam over heavy loam over clay (a) would possess a profile texture value (*V*) of 490.

Depth × texture rating	D × T = V
10 × 16	160
10 × 18	180
10 × 15	150
	——
	490
	——

Where soils are shallower than 30 in and have been derived from the *in situ* disintegration and weathering of underlying limestone rock it has been found, solely by experience, that some allowance should be made if the 'rock' horizon consists of a mixture of rubble-rock and soil particles. The rating for this horizon is taken as one-third of the texture rating of the soil matrix and is calculated to the arbitrary 30 in depth of the standard profile. No such compensation is needed if the shallow soil overlies solid rock pavement.

The estimation of the drainage factor (G). The field diagnosis of the adequacy or otherwise of drainage is based on the criteria discussed under 'Soil drainage' in Item IX, Chapter 2, but is here limited to two categories only, viz.

(a) perfect drainage (well-drained), no gley present,
(b) impeded drainage (ill-drained), gley present.

If the drainage of the soil mass is 'perfect' it is marked as unity. 'Impeded drainage' is evaluated as a decimal of unity on the depth in inches at which true gley appears.

True gley at	9–12 in	Drainage factor 0·5
,, ,,	13–15 in	,, ,, 0·6
,, ,,	16–18 in	,, ,, 0·7
,, ,,	19–24 in	,, ,, 0·8
,, ,,	25–30 in	,, ,, 0·9
Perfect drainage		,, ,, 1·0

The author's experience has shown that gleying at a depth shallower than 9 in does not occur in ordinary arable fields, so that the factor 0·5 is considered to be the lowest limit for the scale compatible with reasonable arable practice.

The profile value. The final value for the soil profile is assessed from the formula:

Texture value (V) × drainage factor (G) = profile value.

If the example quoted above be now supposed to possess a gleyed zone at 27 in, the profile value becomes

$$490 \times 0·9 = 441.$$

The following table shows the order of the soil quality classes:

Profile value	Yields of wheat ears in cwt/acre (from field trials)	Quality rating
600	66	First
500	55 ⎫	
400	44 ⎬	Second
300	33 ⎭	
200	22 ⎫	Third
100	11 ⎭	

In addition to those few methods of survey that have just been discussed, many others could be quoted to the extent of doubling the size of this publication. In almost every country there exists some specific system of soil study either for the reclamation of naturally unproductive soils into a state of valuable productivity or for the more modern utilization of already productive soils.

Well, whatever bit of a wise man's work is honestly and benevolently done, that bit is his book, or his piece of art. It is mixed always with evil fragments ill done, redundant, affected work. But if you read rightly, you will easily discover the true bits and those *are* the book.[1]

If therefore the writer has contributed anything towards helping the natural philosopher to perceive, observe, and record what there is in the soil and its environment to be seen, this little work will have served its writer's purpose.

[1] John Ruskin, *Sesame and Lilies.*

References

AFANASIEV, J. N. (1927) *The classification problem in Russian soil science.* Moscow.

AMERICAN SOCIETY OF PHOTOGRAMMETRY (1952) *A manual of photogrammetry*, 2nd edn. Washington.

BLANCK, E. (1931) *Handbuch der Bodenlehre*, Vol. 7. Berlin.

BORNEBUSCH, C. H. and HEIBERG, S. O. (1936) Nomenclature of forest humus layers. *Trans. 3rd int. Congr. Soil Sci.*, **3**, 260.

BOURNE, R. (1931) Regional survey and its relation to stocktaking of the agricultural and forest resources of the British empire. *Oxf. For. Mem.* No. 13.

BROWN, J. M. B. (1953) *Studies of British beechwoods.* Forestry Commission Bulletin No. 20. H.M.S.O.

BUTLER, B. E. (1955) A system for the description of soil structure and consistence in the field. *J. Aust. Inst. agric. Sci.* **21**, 239.

CHARTER, C. F. (1949) Methods of soil survey in use in the Gold Coast. Conference Africaine du Sols, Goma, 1948. *Bull agric. Congo Belge* **40**, 109.

CLARKE, G. R. (1924) Soil acidity and its relation to the production of nitrate and ammonia in woodland soils. *Oxf. For. Mem.* No. 2.

——(1933) The Brown Forest soils of England. *Forestry* **7**, 43.

——(1951) The evaluation of soils and the definition of quality classes from studies of the physical properties of the soil profile in the field. *J. Soil Sci.* **2**, 50.

——(1954) Soils. *In* A. F. MARTIN *and* R. W. STEEL (eds.), *The Oxford region*, p. 50. Oxford.

——(1962) The preparation and preservation of soil monoliths of thin section. *J. Soil Sci.* **13**, 18.

COMMONWEALTH SCIENTIFIC AND INDUSTRIAL RESEARCH ORGANIZATION (1960) *The lands and pastoral resources of the north Kimberley area, W. Australia.* Land Research Series No. 4. Melbourne.

CORBET, A. S. (1935) *Biological processes in tropical soils.* Cambridge University Press, London.

DOKUCHAIEV, V. V. (1879) Preliminary report on the investigation of the south-eastern part of the chernozem of Russia. *Trudy Imp. Vol. Ekonom. Obs-ches.* Quoted by J. S. JOFFE (1936) *Pedology.* New Brunswick.

140

FRASER, G. K. (1933) *Studies of Scottish moorlands in relation to tree growth.* Forestry Commission Bulletin No. 15. H.M.S.O. Also by the same author: *Peat deposits of Scotland,* part 1. Wartime Pamphlet No. 36. Geological Survey of Great Britain. London (1943); and *Classification and nomenclature of peat and peat deposits.* International Peat Symposium. Dublin (1954).

GLENTWORTH, R. (1954) *The soils of the country round Banff, Huntly, and Turriff.* Memoirs of the Soil Survey of Great Britain (Scotland). H.M.S.O.

GUNN, R. H. (1955) The use of aerial photography in soil survey and mapping in the Sudan. *Soils Fertil.* **18,** 104.

HESSE, P. R. (1955) A chemical and physical study of the soils of termite mounds in East Africa. *J. Ecol.* **43,** 449.

HESSELMAN, H. (1926) Studier över Barrskogens Humustäcke. *Medd. Skogförsökanst. Stockh.* **22,** 169 (with German summary).

HOLMES, A. (1944) *Principles of physical geology.* London.

KRASIUK, A. A. (1929) *Soils and their investigation in nature,* 2nd edn. Leningrad and Moscow.

KUBIENA, W. L. (1953) *The soils of Europe.* Murby.

MATTSON, S. and EKMAN, P. (1935) The reaction and the buffer capacity of soil organic matter. *Trans. 3rd int. Congr. Soil Sci.* **1,** 374.

MILNE, G. (1935) Composite units for the mapping of complex soil associations. *Trans. 3rd int. Congr. Soil Sci.* **1,** 345.

—— (1947) A soil reconnaissance journey through parts of Tanganyika Territory. East African Agricultural Research Station, Amani (1936, unpublished). Reprinted in *J. Ecol.* **35,** 192.

MORISON, C. G. T. and CLARKE, G. R. (1928) Some problems in forest soils. *Forestry* **2,** 14.

MORLEY DAVIES, W. and OWEN, G. (1934) Soil survey of north Shropshire. *Emp. J. exp. Agric.* **2,** 178, 359.

MUIR, A. (1934) The soils of Teindland State Forest. A preliminary survey. *Forestry* **8,** 25.

NYE, P. H. (1954, 1955) Some soil-forming processes in the humid tropics. *J. Soil Sci.* **5,** 7; **6,** 51, 63.

RODE, A. A. (1930) *An excursion to the Lisino experimental forest of the Leningrad Technical Academy of Forestry.* Dokuchaiev Institute of Soil Science, Leningrad.

ROMELL, L. G. and HEIBERG, S. O. (1934) Types of humus layer in the forests of north-eastern United States. *Ecology* **12,** 567.

SANDERS, R. G. (1941) Military cameras for high-speed airplanes. *Photogramm. Engng* **7,** 60.

SOIL SURVEY OF GREAT BRITAIN (1960) *Field handbook.*

SOKOLOVSKY, A. N. (1933) *Problems of soil structure.* Moscow.

SPURR, S. H. (1948) *Aerial photographs in forestry.* New York.

SWYNNERTON, C. F. M. (1936) The tsetse flies of East Africa; a first study of their ecology with a view to their control. *Trans. R. ent. Soc. Lond.* **84,** 31.

TAMM, O. (1920) *Medd. Skogförsökanst. Stockh.* **17,** 49.

TANSLEY, A. G. (1911) *Types of British vegetation.* London.

—— (1930) Use and abuse of vegetational concepts and terms. *Ecology* **16,** 3.

TOPHAM, P. *and* TOWNSEND, R. G. R. (1937) *Forestry and soil conservation in Nyasaland.* Imperial Forestry Institute Paper No. 5. Oxford.

UNITED STATES DEPARTMENT OF AGRICULTURE (1951) *Soil survey manual,* Agriculture handbook No. 18. Washington.

VAGELER, P. (1933) *An introduction to tropical soils* (trans. H. GREENE). London.

VINE, H., WESTON, V. J., MONTGOMERY, R. F., SMYTH, A. J., and MOSS, R. P. (1954) Progress of soil surveys in south-western Nigeria. *C.R. 2me Conf. interafric. Sols,* Léopoldville, **1,** 211.

VYSOTZKY, G. N. (1905) Gley. *Pochvovediniye* **7,** 291.

ZAKHAROV (1927) *Russian pedological Investigations* No. 2.

Index

PHOTO AND MAP SCALES

R/F	INCHES PER MILE	MILES PER INCH	ACRES PER SQUARE INCH	SQUARE INCHES PER ACRE	SQUARE MILES PER SQUARE INCH
1:5000	12·67	0·079	3·986	0·251	0·0062
1:6000	10·56	0·095	5·739	0·174	0·0090
1:7920	8·00	0·125	10·000	0·100	0·0156
1:8000	7·92	0·126	10·203	0·098	0·0159
1:9600	6·60	0·152	14·692	0·068	0·0230
1:10 000	6·336	0·158	15·942	0·063	0·0249
1:12 000	5·280	0·189	22·957	0·044	0·0359
1:15 000	4·224	0·237	35·870	0·028	0·0560
1:15 840	4·000	0·250	40·000	0·025	0·0625
1:19 200	3·300	0·303	58·770	0·017	0·0918
1:20 000	3·168	0·316	63·769	0·016	0·0996
1:21 120	3·000	0·333	71·111	0·014	0·1111
1:24 000	2·640	0·379	91·827	0·011	0·1435
1:25 000	2·534	0·395	99·639	0·010	0·1557
1:31 680	2·000	0·500	160·000	0·006	0·2500
1:48 000	1·320	0·758	367·309	0·003	0·5739
1:62 500	1·014	0·986	622·744	0·0016	0·9730
1:63 360	1·000	1·000	640·000	0·0016	1·00
1:100 000	0·634	1·578	1594·225	0·0006	2·49
1:125 000	0·507	1·973	2490·980	0·0004	3·89
1:126 720	0·500	2·000	2560·000	0·0004	4·00
1:250 000	0·253	3·946	9963·907	0·0001	15·57
1:253 440	0·250	4·000	10 244·202	0·0001	16·00
1:500 000	0·127	7·891	39 855·627	$0{\cdot}^{4}25$	62·27
1:750 000	0·084	11·837	89 675·161	$0{\cdot}^{4}11$	140·12
1:1000 000	0·063	15·783	159 422·507	$0{\cdot}^{5}62$	249·10